PLAYING ODD

Emergence, Intelligence, and the Hidden Logic of Living Systems

Joshua Young, M.D.

Ontic Mind Press

Published by Ontic Mind Press

First published 2026
ISBN: 979-8-9954236-0-7 (eBook)
ISBN: 979-8-9954236-1-4 (Paperback)

Kindle Direct Publishing edition

For Linda:

as complex as the hive, as sweet as its honey

Also by Joshua Young, M.D.

Playing Odd

A podcast on complexity science

Available wherever you listen to podcasts

Table of Contents

Chapter 1: A Structure Too Complex

On September 12, 2023, the following press release was issued, "Human Brain Project celebrates successful conclusion." Begun a decade before, on October 1, 2013, the HBP was an effort to simulate a human brain using supercomputers. The HBP would run for 10 years, as a flagship of the European Commission's newly formed "Future and Emerging Technologies (FET)" initiative. Spread among 155 cooperating institutions from 19 countries, the project was led by the prolific Henry Markram, a South African-born Israeli neuroscientist, professor at the École Polytechnique Fédérale de Lausanne (EPFL) in Switzerland and director of the Blue Brain Project, a 2005 endeavor that sought to create a digital reconstruction and simulation of a mouse brain. To build this simulated brain, the European Union awarded Markram a $1.3 billion grant spread over 10 years.

In 2009, Markram had given a TED talk that ended with the following. "We are going to try and build a brain in 10 years." He suggested that the simulated brain might even attain consciousness, saying "We can do it within 10 years, and if we do succeed, we will send to TED, in 10 years, a hologram to talk to you."

But behind the celebration of its "successful conclusion" and its lists of achievements and accolades, the project had fallen well short of its original goals. Simulating the brain, it turned out, was a significantly more difficult task than had been first imagined. Within a year of its launch, concerns arose about the diversion of resources from other science budgets. These culminated in a letter signed by more than 800 neuroscientists, expressing concerns about the HBP's governance and scientific scope. Under pressure, the HBP underwent a significant restructuring. Its scope was narrowed, from building a complete brain simulation to focusing on more achievable goals, such as developing brain-inspired computing technologies. The brain, the whole brain was, and is, simply too complex.

The adjective we commonly use to describe a truly large number is "astronomical" and for good reason. There are a *lot* of stars. On a clear night, and with good vision, we can see about 5,000 stars, only a tiny fraction of our galaxy's 100 billion. 100 billion is undeniably a large number but, as humans, we have exceeded even this. In 2022 Micron Technology launched a 232-layer V NAND chip containing 5.3 trillion transistors, each connected to the larger integrated circuit by three wires of only three nanometers width, for a total of almost 16 trillion connections. Words about this scale begin to lose meaning to those of us unused to dealing with numbers in this realm. And yet, this unimaginably large number is only about a tenth of the number of synapses, or connections, in a single human brain. The brain is enormous, a genuine galaxy unto itself.

And the synapses of the brain are themselves far more complicated than computer circuits. The transistors that are the elemental unit of a computer's integrated circuit can exhibit one of only two states: on and off, or, in the language of computer science, 1 or 0. The values are discrete, uniform, and repeatable. They are digital.

The neuron and its multiple synapses are nothing like this. The connections between brain nerve cells, or neurons, are what we mean by synapses, and they are roughly analogous to transistors in our example. Each neuron can have as many as 7,000 of these synapses, connecting each of these cells to 7,000 neighbors. But in stark contrast to the uniform and discrete connections of the integrated circuit, different synapses can exhibit different degrees of influence on the state of the neuron. The system is intrinsically analog, and, on top of all of this, is not even static. That is to say, the influence that a particular synapse has on a particular neuron can, itself, change over time. We call this influence the "synaptic weight". The synaptic weights, the connections, are strengthened and weakened as a result of reinforcement, and this is the mechanistic basis of learning in the brain. Synaptic strength and neurotransmitter dynamics add an additional layer of subtlety, of computation, to the dynamics of the brain.

It goes without saying that these circuits and these synaptic weights are not random, that they serve precise functions and demand a complex architecture. But from where does this architecture come? Where is the blueprint?

Six years before Henry Markram's TED talk, another science megaproject reached its culmination. The Human Genome Project received $3 billion in funding from the National Institutes of Health and the Department of Defense with the objective of sequencing, that is to say mapping, the entire human genome, the collection of genetic material of our combined 46 chromosomes. In the event, the project sought to map a human haplotype, or half of our paired 23 chromosomes, for a representation of the entire genome within 15 years. At its conclusion in 2003, 92% of the genome had been mapped, although the remaining 8% would take an additional 18 years of work.

Described to the public as the "book of life", our genome is composed of "coding regions" or genes and much larger stretches of non-coding regions, what used to be called junk DNA. Prior to the launch of the Human Genome Project, the consensus among scientists was that the human genome contained about 100,000 genes. This "book of life", the construction manual for an entire human, is remarkably small. Data are contained in nucleotide base pairs, or simply *base pairs*, analogous to bytes in a computer program. The combined genetic material of all 46 chromosomes amounts to over 3 billion base pairs or 770 megabytes of data. That is 770 megabytes, not terabytes or even gigabytes. 770 MB would hardly make a dent in the smallest USB drive. Of these 770 MB, only about 2% is coding; only 2% are actually genes. This hardly seems enough to build a brain, let alone an entire human being.

At its conclusion, the Human Genome Project found that the estimate of the total number of genes in the human genome was wrong. There were not, in fact, 100,000 genes to build a human. There were far fewer, only about 20,000-25,000.

This remarkable finding of the human genome project, of the paucity of human genes, makes our question about blueprints even more salient. How can 20,000 genes encode for 100 trillion synapses?

Here, we come to one of the major themes of this book: the idea of information compression. We will see compression come up in our discussions of simulation and, indeed, of understanding itself.

Those of us who use computers, which is to say all of us, employ data compression every time we zip a file. The utility of doing so is obvious in that compression allows us to send large files and photos that would otherwise exceed email constraints. For purposes of illustration, I downloaded the UK's National Archives' modern English translation of the Magna Carta, a proto constitution in which King John ceded some rights to those of common birth. The text runs to just over 4,500 words. Saving this text as individual letters in txt format, the file comprises 25,000 bytes of information. Compressing these 25,000 bytes of characters, spaces, and punctuation as a zip file reduces its size to 10,000 bytes, a reduction of 60%. This sort of compression is called lossless because the original information can be reconstituted with perfect fidelity. We can achieve much greater degrees of compression by applying lossy strategies. When we save photos as jpegs and music as MP3, the algorithms actually remove some of the original data to compact the file not 60% but by 90% or more. However, the reconstituted JPEG picture or MP3 song does not exactly match the original. Depending upon the chosen degree of compression, often termed "quality" in computer lingo, the difference may not be discernable to the listener, but the reconstituted file is not completely faithful to the original.

Genes do not lend themselves to lossy compression because the substitution of even a single nucleotide can result in a dysfunctional product. Indeed, sickle cell disease is attributable to such a single base pair substitution. The discussion of single nucleotide polymorphisms is involved and beyond the scope of this book, but the fact remains that genetic compression is limited to the lossless.

Genetic material, in some form, is thought to have been extant for more than 3.5 billion years, the estimated age of the Last Universal Common Ancestor or LUCA. That is a lot of time for natural selection to foster genetic compression strategies and indeed our genome exhibits many. These can be broadly subdivided into regulatory elements and alternative splicing. Regulatory elements are parts of the genome that encode not for protein synthesis but rather act as controls to turn genes off and on. They are like the conditional statements in a computer program, the if/then/else statements. I imagine many readers will be familiar with computer programming, but I wish to include those who are not and so will give a non-programming example.

In the stock market, itself a complex system to which we will return, it is common to place conditional trades. These take the form of "If stock X hits price Y, then sell Z shares" or buy, as the case may be. For example, a conditional order may be given to a broker to buy 1,000 shares of Google if the share price dips below $120 and buy 5,000 shares if the price dips below $100. This straightforward order is mostly composed of conditional statements. The only action command, analogous to gene activation, is "buy 1,000 shares" and "buy 5,000 shares". The remaining nearly 70% of the order is composed of control statements, of regulatory elements. Such is true of the genome, only to a much greater extent. While only 2% of the genome codes for the action of producing proteins, the remaining 98% are not devoid of information. Perhaps a better way to envision the genome is as being composed of two sorts of information: information about product and information about process. The gene contains information directly related to its protein product; the regulatory regions contain information about the process of evoking and suppressing production. We will see that this product/process dichotomy is a defining feature of complex systems.

As humans, we are unused to this way of thinking. The blueprint of my office describes the doors and walls, but not when to build them.

The design of the cabinetry is just that, a design of a thing. It is not an instruction to build a shelf only when fewer than 3 shelves lie beneath it. To truly understand complex systems, we need to see the interplay between process and product and much more besides.

In this chapter, we have come from marveling at the immensity of the brain and seen our hubris at trying to encapsulate it in computer hardware, to our surprise at the tininess of our genetic blueprint, to strategies of expanding the effect of this genetic information through regulation and conditional activation. But do these genetic controls provide enough information to build a brain? In our exploration, we'll see incredible genetic and epigenetic tools, talk about non-genetic information, and ask what we mean by "information" in the first place. In the meantime, consider your own brain, vaster than a galaxy and built out of so little.

Chapter 2: Doing More with Less

On August 6, 1940, Nikolai Vavilov was arrested and subsequently sentenced to death. His crime: being a geneticist. For the Soviet Union in general and Joseph Stalin in particular, 1940 was a busy year. One year after the signing of the Molotov-Ribbentrop non-aggression pact between Nazi Germany and the Soviet Union, the Nazis had invaded France and embarked upon the creation of a Tripartite pact with Japan and Italy, what we know as the Axis Powers. Stalin had again sent Molotov to the negotiations to see whether the Soviet Union could become a permanent partner in this alliance. Stalin was busy. But not so busy, it seems, for him not to personally involve himself in the affairs of a Soviet agronomist, Trofim Lysenko, even going so far as to later personally edit a speech that Lysenko would subsequently give.

Lysenko's thesis, echoing the pre-Mendelian inheritance theories of Lamarck, was that events in a person's life would themselves influence traits passed on to his children. The example commonly given is the expectation that the blacksmith's son would have well-developed biceps because of the exercise the father had done. This idea of behavior and environmental exposure resulting in attributes that are heritable and are passed on to descendants was very much in accord with Soviet political theory, particularly the idea of the New Soviet Man, as an archetype arising from a post-scarcity communist society and with heightened traits that would be passed on to children for generations to come. Lysenkoism was central to what became known as Stalinist Science or Marxist Genetics, an amalgam of scientific and political theory.

As a young man, Lysenko received encouragement from the better-known Nikolai Vavilov, a geneticist and botanist and proponent of what was known as Mendelian genetics, the idea that genes determine inheritance, that both parents contribute to inheritable traits, and that some genes are dominant and some recessive. Mendelian genetics is the idea behind big B little b brown eyes versus blue eyes genes. This mechanistic and unmodifiable determinant of physical attributes was anathema to Stalinist political theory.

Vavilov might have been ignored had he not gone out of his way to denounce Lysenko. Lysenko, for his part, denigrated Mendelian genetics as bourgeois science. And apparently, this was sufficient cause for Nikolai Vavilov's arrest and subsequent death sentence. Objective science, that is to say the truth, had been suborned to political ambition.

And here is the irony. Decades later, Lysenko's ideas of the life experience of an individual resulting in traits that are passed on to progeny, the very idea that Stalinist science promoted over empirical evidence, that idea turns out to be at least partly true.

In Chapter 1, we discovered the almost unfathomable complexity of the human brain composed, as it is, of more than 100 billion synaptic connections. We were surprised by the finding of the human genome project that the sum of human genetic material comprises only 20,000 to 25,000 genes and we wondered how it is possible for this sparse genetic information to produce the precision and scale of the computational circuitry of the brain, the human *connectome.*

We explored the idea of compression of genetic information and the presence of extra-genetic regulatory DNA that affects the expression of genes, allowing us to squeeze more information out of a limited resource. We introduced the idea of the product/process dichotomy in the genetic code and emphasized that this will be a theme that will appear on multiple occasions throughout our discussion. But for all that, we still remained impossibly far from understanding how to get from 25,000 genes to a structure more than six orders of magnitude larger. And it is with this question that we begin this chapter.

It is patently apparent that environmental exposure can affect gene expression. Every time we get a cold or are vaccinated, we are exposed to viral protein that elicits an immune response, the production of antibodies to the particular proteins of the germ. What is this antibody production other than triggered gene expression? Of course, we do not expect our vaccinations to result in immunity that is passed on to our children, and indeed it is not. However, there are some exposures that seem to result in a chemical change to selected genes themselves. The most well-known of these is methylation.

The methyl chemical group is very simple. Unattached to a larger molecule, CH_4 is known as methane, or natural gas. Unlike DNA, there aren't any configuration options to the methyl group. The DNA molecules of your chromosomes differ from each other and from those of other people, but every methyl group in your body is the same. The methyl group doesn't contain any information itself, except about its presence and location. Instead, the methyl group serves as a flag.

Both genes and histones, the structures around which DNA is wrapped, can be methylated. These methyl groups cause certain genes to be more available for expression, that is to say for protein synthesis. The methyl groups do not change the information content of the gene, nor do they add any information themselves, they simply make it more or less likely that a particular gene will be expressed. There are two good ways to think about methyl groups. They can be seen as switches on a railroad track that can determine whether a train will pass down the line or be shunted to a siding. In this respect, they are highly influential. However, the railroad switch does not alter the contents of the cars of the train. They influence expression, but they do not add any information to the gene itself.

Another good way to think of methyl groups is as punctuation. The gene itself is a sentence composed of elements that together form the instructions for the assembly of a protein. The methyl groups may represent commas or semicolons or periods that terminate the expression of the gene. They are very important, but they do not add to the information content of the sentence.

If they do not add much information, other than their presence and location, why even discuss them in the context of neurodevelopment? It is because methylation may occur from an exogenous influence. That is to say, a gene or histone may become methylated because of environmental exposure. Indeed, the developmental abnormalities associated with diethyl stilbestrol (DES) seem to be associated with changes in methylation of DNA. Bisphenol A or BPA, a chemical found in some plastic bottles, may also alter methylation and may even affect obesity. Even exercise has been associated with epigenetic changes.

We generally think of genes of DNA as being the sole carrier of traits from parents to children. But if methylation, an epigenetic feature, can affect the development of a child, can it too be passed on from generation to generation?

This turns out to be a trickier question than it initially appears because there are two sorts of multi-generational traits, intergenerational transmission and transgenerational transmission. The proof of the pudding seems to be in the third generation. I will explain.

In the case of DES, the mother is exposed to the chemical and it is not impossible that the mother will also develop epigenetic changes even if these do not manifest in any obvious way. The fetus she is carrying undoubtedly undergoes epigenetic changes in the manner described above. Therefore, the single exposure has affected 2 generations: the mother and the fetus. But what if the chemical also alters the cells in the fetus that will go on to become egg or sperm. This sort of effect, resulting from a single exposure, would then affect not only the mother and the fetus but also the eventual children that the fetus might have. In this way, single exposure would affect 3 generations: the mother (called generation 0), the fetus (called generation 1) and the fetus's children (generation 2). In order to establish that a trait has actually been inherited rather than resulting from a single multi-generational exposure, the trait must be present in the *third* generation, in generation 3, that is to say the great-grandchildren of the mother who is exposed. Incredibly, such demonstrations have been made, especially in plants in which methylation itself is replicated and passed on to progeny.

The fidelity of replication of DNA is astounding. Copying errors occur, but they are impressively infrequent. The number of base pairs that are incorrectly copied in DNA replication is only about 1 in 10 million to 1 in 100 million. At least in plants, an enzyme exists to copy methylation marks so that they are replicated along with the DNA. These enzymes, methyltransferases, demonstrate a much lower degree of fidelity, with errors occurring once in every 25 methylations. Still, the effects of these epigenetic changes are, for the most part, carried across many generations.

Although there is some suggestion of transgenerational epigenetic transmission in mammals, it is at least far less likely than in plants, and for a very interesting reason. Methylation serves a function in embryo development that is quite apart from an environmental exposure. Methylation and other epigenetic changes are an important mechanism for cell differentiation. All the cells of our body, with the exception of red blood cells, contain the same DNA (although sperm and egg cells contain only one copy rather than two). The question then arises, if brain cells and kidney cells have the same genetic content, why are they different from each other? Methylation, primarily of histones, varies by cell type so that even though the data for production of every protein is available in each cell, the relative expression of these data varies as a result of epigenetic changes. A kidney cell and a brain cell may have the same DNA, but they will not have the same methylation in the same places.

On reflection, this is a bit of a problem because two particularly specialized cells are egg cells and sperm cells. These cells have also differentiated from the original cell line of the embryo from which they arise. But upon fertilization, they must become undifferentiated so that they can serve as the new seed from which all else will grow.

Cells that are capable of differentiating and cells that are themselves undifferentiated are referred to as stem cells because they are the stem from which other cells will form. Stem cells can be pluripotential, meaning capable of turning into, of differentiating into, many different sorts of cells, or totipotential which means so undifferentiated that they are capable of transformation into any sort of cell. For obvious reasons, the fertilized egg must become totipotential because it is from this fertilized egg that an entire organism must derive. Since epigenetic changes like methylation are markers of differentiation, these must be erased in order to reset the fertilized egg into a totipotential cell that can divide and differentiate into every cell from which the organism is composed.

This erasure of methylation marks, of epigenetic changes, makes it impossible to pass methylation on to the next generation. Well, not quite impossible it seems. The erasure is often incomplete and there is evidence that some of these epigenetic marks do in fact pass through trans-generationally, to progeny through inheritance rather than exposure. Methylation marks that fail to be erased are often termed *escapees.* The study of methylation escapees in mammals is complicated and largely ambiguous with few concrete examples. One example, in mice, is the evidence for transmission of obesity.

In order to make transgenerational detection a bit easier, patrilineal, that is to say transmission through the male line, is studied. The zero generation, that is the original father mouse, may have environmentally induced methylation changes. The sperm of that mouse, necessarily contained within the same body as the father, may also undergo methylation changes from this exposure, but the intergenerational transmission ends there. The grandchildren of the original father mouse, if they demonstrate the methylation-associated trait, represent evidence of true transgenerational transmission. There is evidence that dietary restriction or other dietary changes may result in methylation changes that are passed on for several generations.

This seems to be the exception rather than the rule in mammals and the evidence is even weaker for transgenerational epigenetic transmission in humans. Although dietary changes, exercise, and emotional trauma have been suggested as associated with methylation, studying the effects in later generations is difficult because other aspects, completely independent of genes, are also passed on generationally. Factors like poverty, education, and access to resources are also intergenerational and can affect the development of children. As the biologist Stephen Jay Gould wrote in his book, The Panda's Thumb, "Human cultural evolution, in strong opposition to our biological history, is Lamarckian in character. What we learn in one generation, we transmit directly by teaching and writing."

Although fascinating, we have to ask what the relevance is in addressing our project of identification of sources of information for neurodevelopment. That is to say, if the genome itself is inadequate to produce such a complex structure, can the information arise from elsewhere? Even though methyl groups do not carry much information, the information that they do carry may arise from the external world. This small amount of information does not come from the organism itself. To the extent that it has any influence on neurodevelopment, the information, the data, arise from the outside world.

Although the amount of data conferred from outside of the organism is indeed very small, we will see the theme of "exogenous information" recur at a significantly larger scale much later on.

Let's ask a more fundamental question. Can the expression of a gene, even if it occurs multiple times, produce a structure that is more complicated than a single expression? In fact, let's state this query more basically: can a set of instructions, if repeated, produce a structure more complicated simply through repetition? We will answer this question in the next chapter.

In this chapter we asked whether factors outside of our genes can influence development and whether this influence is inheritable. We discovered that environmental influences in the parent organism are passed to many generations of progeny in plants and that some evidence suggests that limited transmission may occur in mammals as well. Looking more broadly at the flow of information that directs the development of an organism, we saw that environmental factors represent "exogenous information" and saw that the instructions for building an organism are only partly contained within the organism itself.

Chapter 3: The Structure of Noise

International Business Machines, IBM, developed a product that defined an entire market segment, capturing an incredible 94% of the share before technological progress made it obsolete. There was no office of any worth that did not have several of these machines. In 1981, IBM introduced the personal computer, but this is not the product of which I speak. The IBM PC only ever captured about 80% of the market share and declined rapidly over the course of the following years. No. Twenty years earlier, in 1961, IBM introduced an iconic product, the IBM Selectric, a typewriter that replaced the type bar with a type ball that would rotate to the correct character almost instantaneously. Unlike conventional typewriters, the ball of this electric typewriter could be easily swapped with a ball of a different font.

Beyond being an excellent typewriter, the Selectric also interfaced with early computers to serve as a printer. If you've never seen one of these marvels, I invite you to hop over to PlayingOdd.com where you will find links to IBM's historical web page and also to the opening credits of the 1970s British science fiction television series called UFO that established its futuristic bona fides in the opening credits by showing a Selectric printing out an important message.

The system was excellent, but not perfect, and the Selectric would occasionally print out an incorrect character when used as a printer. These typographical errors were caused by noise in the transmission line between the computer and the Selectric. Indeed, communication between computers, often conducted through phone lines, was plagued with noise and this noise, obscuring the data signal, set a limit on the amount of information that could be transferred. This was a substantial problem for IBM whose business was predicated on transmission of data.

In the 1950s and early 1960s, IBM sought to address the noise problem by improving the components used in the system, by increasing the strength or amplitude of the data signal, and by introducing low pass signal filters. A low pass filter does exactly what you would expect, it lets lower frequencies pass through while removing higher frequency signals from the transmission line. Since data were generally passed in lower frequency signals, the higher frequencies tended to disproportionally represent noise.

This approach yielded some initial benefits, but despite impressive engineering, IBM simply could not get the noise out of the system. A different approach was needed and for this, IBM brought on a Polish polymath, Benoit Mandelbrot. Born in Warsaw, Mandelbrot fled to Paris in 1936 on the eve of the Second World War. As a Jew in Nazi occupied France, Mandelbrot lived in fear of denunciation by a rival and arrest by the Nazis, as had happened to a close physician friend of his, Zina Morhange.

Paralleling Albert Einstein, another Jew who had fled from the Nazis, Mandelbrot joined the faculty at the Institute for Advanced Study in Princeton NJ where he was sponsored by John von Neumann. In 1958, having been recruited by IBM at the Watson Research Center in Yorktown Heights NY, Mandelbrot was assigned to the problem of intractable noise in computer communication.

Eschewing the approaches that IBM engineers had taken to suppress noise, Mandelbrot recognized the noise in the signal as something more intrinsic. His insight was to view the noise over different scales of time. To understand his great contribution, we need to see noise as something visual. Let's imagine a graph in which the Y axis is the amplitude of the signal and the X axis is time. I deal with graphs like this all the time when recording podcasts, in which my voice modulates the height of the Y axis and time proceeds in seconds along the X axis. But even when I do not speak, there is some bumpiness in the graph that forms. Some of this bumpiness arises from sounds from outside of my recording studio. Some of the bumpiness arises from random electromagnetic fluctuations in the cable between my microphone and preamplifier and between the preamplifier and my computer.

Mandelbrot's great insight was to compare the bumpiness of this signal, that is to say the noise, over different scales of time. He found that whether the scale of the X axis was in large or small units of time, the graph of the noise looked the same. He referred to this property as self-similarity, in my example across time.

But let's imagine this a different way. Imagine that we capture 10 minutes of random noise. Regardless of the degree to which we zoom in on this plot, the pattern of the noise will look essentially the same, although, of course, not identical. In fact, if we were simply presented with a graph of the noise with an unlabeled X or time axis, we would not be able to surmise whether we are looking at 10 minutes of data, 10 seconds, or 10 milliseconds. This *self-similarity* over scale is a fundamental characteristic of what became known as fractal geometry. A fractal is a shape that looks similar regardless of how much magnification is employed in its observation. Self-similarity is a feature of many biological systems. Think of the branches of the tree. If one were to make a line drawing of the branches, it would be impossible to tell whether the drawing is of branches coming off a trunk or of twigs coming off a smaller branch. The pattern, although not identical, is roughly the same.

The fractal nature of many biological structures is observed in the bronchioles of the lungs, the vascular branching of blood vessels in the body, and perhaps most iconically in the beautiful spirals of Romanesco broccoli. Changes in the range of scales over which a biological system is self-similar may themselves be indicators of pathology, and this is the subject of some of my own work.

Mandelbrot showed us that fractal geometry, patterns demonstrating self-similarity over a range of scales, can be used to describe natural phenomena like noise, coastlines, mountain ranges, and many biological systems. But it is also possible to take the opposite approach. That is to write a simple set of rules that will create a pattern that looks intrinsically biological.

The example that I will give is not so much of a biological pattern as of a recognizably natural one, but it is a pattern that lends itself to an experiment you can perform with ordinary household items. For this experiment, you will initially need four matches or other similarly shaped objects. Let's line up three of those objects into a line that is three matches long. Now I'm going to introduce a rule. The rule is that the middle 1/3 of the line, that the match that makes up the middle 1/3 of the line, is replaced by two matches. We are not going to move the match on either side of that middle section. So in order to keep all of the matches connected, the two new matches must be tented up to form a point. Now we have four matches: the ones on either side being unchanged from the original pattern and two matches in the center that form a point.

So far so good? OK. Now we're going to do the same process with the same set of rules but rather than starting out with a line of three matches, we are going to start out with the triangle. Each of the sides of the triangle is composed of three matches to make a line just like the one we started out with. Therefore, the triangle as a whole will be composed of nine matches, three for each side. This looks like an entirely unremarkable equilateral triangle.

We are now going to introduce our same rule. The middle match of each side is replaced by two matches forming a point. If we do this properly, we will have transformed our triangle into a six-pointed star. If you've gotten this far, bravo. You are almost done with the experiment.

The next part we are going to do will be in our imagination, unless you are really keen on this project. Each of the sides of the star can be thought of as representing three small segments or three mini matches. We are going to, in our imagination, replace the middle 1/3 of each of these with two small segments again forming another point. If we continue to apply this rule of replacement of the center 3rd of each edge with two sections making a point, we will produce, ultimately, an outline that looks very much like a snowflake.

The pattern we have made is intrinsically fractal. I say this because we can zoom in on any of the edges at any scale and the pattern will look the same. Indeed, drawing out the pattern of part of one of the sides of the snowflake, it is impossible for us to guess at which scale we are viewing it because, at all scales, the outline looks the same.

What have we done here? We have taken an extremely simple rule, that of the substitution of the center 3rd of each edge, and used it to produce a pattern of essentially infinite intricacy. A rule that is applied and then reapplied is referred to as being iterative. Iteration is itself an example of a sort of compression. We did not have to create an instruction set that described the complicated outline of the snowflake. We only had to establish a rule and then iterate it. The instruction set is very simple. The same sort of iterative strategy is employed in genetic expression and can generate really very complicated structures.

Could iteration be the secret to generating our fantastically complicated brain? To answer that question, let's compare our snowflake to an actual snowflake. If all snowflakes grew by following one simple rule, they would all be identical. And yet the adage that each snowflake is different turns out to be largely true. From where do these differences arise? As the snowflake is buffeted in its fall to earth, the likelihood of an ice crystal forming is not equal on all its edges. These random outside influences, what we can think of as noise, affect the final shape of the snowflake. And since no two snowflakes encounter exactly the same number of molecules of water in the same orientation at the same times, the final structure of each snowflake is indeed slightly different.

Our iterative rule to construct a beautiful and perfect snowflake is also the greatest problem with this methodology. Because we have a fixed rule, the process of producing the snowflake is what we refer to as determinative. That is to say, we know what the pattern will be because of the rule that is applied, and the outcome must necessarily be the same or else we have made some error in our calculation.

There are biological structures that can function perfectly well even if their structure is determinative. It is unlikely that the patterns of the veins on a leaf are so important that they cannot all be the same.

This is patently untrue of the brain. The connections in the brain are neither random nor entirely determinative, because were they so, each of us would have an identical brain and, indeed, each portion of our brain would be identical to every other portion. Amongst other motifs like the importance of exogenous information, noise will appear several times over these chapters, and will prove itself not always to be an impediment to construction but sometimes a very element of construction itself. Iterative construction will play a role in our unfolding story as a bit part but not as the main character. Ours is a story about information and how much information can be contained in an infinitely branching pattern if the pattern itself is created by a single simple rule. In order to answer this question, we need to define what precisely we mean by information. And that will be the subject of the next chapter.

In this chapter we learned about self-similarity in natural systems. We saw that a simple rule, when repeated, can produce a pattern of infinite intricacy. We observed that rule-based construction of a pattern always yields the same result, and we named this ultimate predictability as "determinative". Finally, we drew a distinction between structures that are intricate and structures that are information laden, although we have yet to define information.

Chapter 4: Information

We begin with the first stanza of a poem written by our hero:

Forty-three quintillion plus

Problems Rubik posed for us.

Numbers of this awesome kind

Boggle even Sagan's mind.

Some chaps pry their cubes apart

In with calm intelligence.

Then reassemble to the "start".

Not cricket! A rude game's afoot

And up with which we will not put!

And, true to form, the poem is complete with references. Any guesses?

How about this? He is famed not only for building a juggling robot but also an electromechanical mouse that *learned* its way through a maze. I want to belabor a point here. The mouse was electromechanical. It was not electronic. It had no RAM and no integrated circuits. It merely used mechanical parts, circuitry, and magnets to exhibit a degree of artificial intelligence.

Juggling, mechanical mouse, optimization of Rubik's Cube solving algorithms to say nothing of his poetry of the same. Still not there?

Here's the final clue. This genius, whose 1937 master's thesis, not PhD dissertation -- his master's thesis -- was described as possibly the most important and also most noted master's thesis of the century, invented and named the concept of the "bit". Our man of the hour is the venerated Claude Shannon.

Our quest from Chapter 1 has been to understand the nature of information required to construct a brain. And yet, it has taken us to Chapter 4 to get around to defining what we mean by information.

For most of us, the idea of information, the definition, is self-evident. Information is something that tells us something we did not previously know. The temperature outside at this very moment represents information that we don't immediately have but can gain by looking at a thermometer or turning on a news radio station. Our ideas about information are not dissimilar to Renaissance ideas of matter. Anchored in Aristotelian thought, the idea was that matter originated as a raw material, a generic "substance". This substance, through maturation or by external means, could be transmogrified into a more specific material, into a more derived material with specific characteristics: gold, grass, cow. The revolution in thought that led to our modern idea was the concept of quantization of matter into atoms, what we refer to as elements.

The nature of information was similarly amorphous until it too was quantized. The idea of an element of information, a base unit, was the invention of Claude Shannon. He defined the smallest unit, the atom of information, as a binary digit, an element whose value could be zero or one. All other information could be built from this binary digit for which he coined the portmanteau word "bit".

Combinations of bits allowed for more than two permutations and 8 bits were sufficient to represent the upper and lower case characters and punctuations of the English alphabet. Our term for a collection of eight bits is, of course, a "byte". When we compare our hard drive sizes, we speak about megabytes and gigabytes and terabytes. Interestingly, and undoubtedly cynically, when we compare broadband speeds, these are generally given in bits to make them seem larger.

The creation of the bit, the unit of information, allowed mathematical comparison of different sorts of information. Is a picture really worth 1000 words? 1000 words is roughly 5200 bytes or 42,000 bits. Although it is possible to produce an image that is only 5K, it would be an awfully small or awfully low resolution image. So, a picture is indeed worth 1000 words, but any decent picture is worth a great deal more. This sort of information comparison would have been impossible without the idea of a unit of information.

The amount of information in a string of letters is not directly dependent on the number of letters in this string. This sounds counterintuitive but will become obvious with an example. The previous sentence, "this sounds counterintuitive, but will become obvious with an example" comprises 70 characters including spaces. One might think that any sentence with 70 characters would require the same number of bits, that is to say, it would contain the same amount of information. But what if those 70 characters were all the letter A, just 70 A's in a row. I could describe that sentence as 70A's, but the phrase "70A's" itself contains only 5 characters. Therefore, the information contained in a string of 70A's is no more than five bytes compared to the 70 bytes of the original sentence. As you might have guessed, a string of 70A's conveys a great deal less information than a sentence might, and now we have the tools to directly compare the two.

What is it about the 70A's that makes it so information poor? The most obvious reason, and indeed the correct one, is that all the elements of this sentence are the same. In more general terms, there is more similarity between elements in the 70A sentence than there is in the original sentence. Similarity within a data set represents a decrease in the information conveyed by those data.

At the end of Chapter 3, we had constructed a snowflake of infinite intricacy through the iteration of a simple algorithm substituting the middle 3rd of each line with two additional segments forming a point. We asked whether this sort of iterative fractal process is sufficient to generate a structure as complicated as the brain, because if it were so, then this sort of iterative genetic impression would be our answer to overcoming the limited size of the genome in brain construction.

Let's take the perspective that Shannon might have in evaluating the amount of information that is added through iteration to produce what we all can agree is an incredibly intricate snowflake. The answer lies in our comparison of elements of the snowflake, analogous to our comparison of characters in the sample sentence and in a string of 70 A's. We found that 70 A's contained very little information in comparison to a sentence of 70 characters because of similarity between elements, that is to say the letters, in the string of 70 A's. Similarly, the snowflake demonstrates tremendous similarity in the outline of its shape. Indeed, self-similarity across scale is the defining characteristic of fractal shapes. It would therefore be possible to compress our description of the entire snowflake into a very small data set with multiple repetitions. The snowflake, although infinitely complicated and with limitless angles and points, contains very little information indeed.

On reflection, this should come as no surprise. The generation of the snowflake required one simple rule repeated an infinite number of times. The shape, therefore, distills down to this simple rule.

This is highly relevant to our project of producing a human brain from a limited instruction set. We can see very clearly the distinction between intricacy and the information density. The organization of veins in a leaf or branches on a tree, of bronchioles in the lung and of venules in the retina, although intricate, contains much less information than one might otherwise expect.

This would not be a problem for us if the brain were simply composed of identical branches of neurons. But it is not. The differences in connections between neurons are precisely what determines brain function and the fact that they are not all the same is evidence that the amount of information required to build a brain is far more than iterative fractal processes can provide. This is not to say that fractal processes will not play a role in our eventual construction of the brain. But this is to observe that iterative processes will not add much information to the final product.

Let me give you another analogy. We can describe the construction of a wall by beginning with a line of bricks with appropriately placed mortar. Our instruction, to be iterated, is to lay the next row of bricks so that their ends overlap with the ends of the bricks below. If we repeat this instruction, we can build a wall of any arbitrary height. We could build a wall much larger than any house that exists today. But we could not build a house. And this is because the structure of a house requires more information than this simple instruction. It is important for us not to confuse arbitrary size and intricacy with complexity.

Noise, both as a confounding and a constructive element will appear many times in this book. Claude Shannon's information theory has a great deal to say about noise, much of it counterintuitive.

Let's return to our exploration of the meaning of information. How much information is contained in the number 185.72? We now have the tools to answer this question. Since the measure of information is bits, we can ask: how many bits does it take to represent 185.72?

Information theory answers this question by asking a question of its own. How many possible values are there? Let's say 185.72 is the stock price of an IBM share. In fact, let us say that 185.72 is yesterday's closing stock price. Did you know that? Let's say that you did not. Then the number 185.72 would indeed convey some information to you. However, it is easy enough to find out what yesterday's IBM's closing stock price was, and let's say that you did know that it was 185.72. Then how much information is conveyed by my giving you yesterday's closing IBM stock price? The answer is none at all. You have no more information than you did before I told you something that you already knew.

But what if 185.72 is tomorrow's closing stock price? This would seem to convey a great deal of information, and indeed it does. Information theory would ask, as it did in the last example, how many possible values are there? Now, there are not infinite stock prices that the IBM share can have, but there are certainly many. Let's say, in practical circumstances, there are 10,000 different values. To determine the correct answer amongst those 10,000 is to convey a great deal of information. To oversimplify things, the amount of information might represent 10,000 bytes. In actual practice, the calculations are done differently, and distributions are taken into account. But even disregarding this, it is obvious both mathematically and intuitively that tomorrow's stock price carries a great deal more information than yesterday's stock price.

Let's try something a bit less intuitive. Claude Shannon died in 2001. Let us say we are interested to learn the year of Claude Shannon's birth. We can safely say that the year of birth must have been sometime between 1878 and 2000, since the oldest recorded human being lived fewer than 123 years and because it is unlikely that Shannon worked all of his marvels as a one year old infant. In the event, Claude Shannon was born in 1916. How much information is contained in the four digits 1916? If the number 1916 refers to the year of Claude Shannon's birth, then indeed some information is conveyed since there were about 120 potential answers from which to choose.

In order to determine the amount of information conveyed, we must necessarily ask from how many possible answers does this data point come? Let's say the question is choose a whole number between 1915 and 1916 inclusive. In this case, the number 1916 conveys only a single bit of information since only two possible answers, 1915 and 1916, exist.

Here's the counterintuitive part. Let's say the question is not in what year Claude Shannon was born, but in what year was Claude Shannon born <u>followed by two random digits</u>. For example, if the two random digits are three and seven, the answer would be 191637. If the random digits were five and six, then the answer would be 191656.

By appending these two random digits to the end of the year of Claude Shannon's birth, have we added information? We can determine the amount of information conveyed by the number of possible answers to the question asked. By appending two random digits to the end of the year of birth, we have multiplied the possible number of answers by 100. Does this mean that this number carries 100 times as much information simply by appending two random digits? The answer, counterintuitively, is yes. There are more potential answers and therefore there is more information carried by the actual answer.

Let's explore this a little bit. From where does this extra information come? Does it come from randomness itself? The answer is yes. How often do questions like this arise? Also counterintuitively, all the time! Let's say that I wish to transmit the number 1916 through a telephone line (after all, Claude Shannon did work for Bell Labs). Although 1916 is all that I want to transmit, there will be some noise in the transmission line because there is always some noise in the transmission line. Since my data are conveyed down this line as electrical signals and since noise represents additional random electrical signals, the situation is entirely analogous to adding random digits onto the end of the data I wish to transmit. This random data represents information. It is not the information I intend for the recipient to receive, but it is information, nonetheless. Indeed, the reason it is so difficult to remove noise from a signal is not because the noise represents no information, but precisely because it represents information that is in addition to the information I am trying to transmit.

In the example of the sentence of the string of A's, we observed that the information density of a message is related to the similarity of the elements within that message. Because each of the letters A in this string of A's is identical to the next, it is possible for us to describe this string of characters more compactly than we could in an actual sentence. Compactness, as you already know, is one of the main motifs of this book, since we intend to so compactify the connectome of the brain as to fit it into the genome, if that is possible. Producing compactness is what we call compression and we have already seen compression arise as a subject of several of these chapters. We can use compression as a measure of the amount of information contained in a message.

You can try this little experiment yourself. Copy a section of text of 1000 characters. Save that file as "sentences.txt". Now create a second text file composed of 1000 letter A's. Save this as "a.txt". Now zip (sometimes called compress) each of these files. Which one of the files compresses to a greater degree? You'll see that the file consisting of letter A's compresses to a much smaller size. This is indicative of low information content.

Now let's do the same thing with a file consisting entirely of random numbers, that is to say, of noise. Obtaining random numbers is surprisingly difficult. When we use Excel or other software to generate random numbers or to use a random function, the numbers are not truly random but are generated by an algorithm to produce numbers that are random-ish. We refer to these as pseudorandom number generators and, for most scientific purposes, these are adequate. But for our exploration of information, they are not. Obtaining true random numbers usually means performing an actual physical measurement on something that is intrinsically random. The classic example is the noise of the cosmic background radiation, the leftover radiation produced by the Big Bang. These numbers are felt to be genuinely random and are downloadable. In fact there are websites that offer truly random numbers and I have put a link to one of these in the reference section. If we copy 1000 characters from a truly random data set, and try to compress or zip the file, we will see that not much compression can be obtained. Noise, true random content, is almost incompressible. It is so information dense that there is no way to accurately summarize it. This is real physical evidence that noise represents maximal information density. To make this more palatable, we should probably draw a distinction between information that we want and information that we don't want. In electronics, information we want is referred to as signal and information we don't want is referred to as noise. But as we have seen, noise actually carries more information than signal, and what we are really trying to do when we attempt to isolate a signal from the noise is to isolate a small information set from a larger information set.

The information content of noise plays a significant role in cryptography, the science of encoding information so that it may not be observed by unauthorized parties. The most secure sort of information encoding is the use of what is called a one-time pad. To try a simplified version of one-time pad encoding, all that you need is a message that you wish to encode and a book. Take the message you wish to encode and look at the first character. Now look at the first character of your book and then add the two where A = 1, E = 5, et cetera up through Z = 26. Translate this back into a letter and then write that down as your encrypted letter. Continue this for your entire message and you will have produced a code that is unbreakable unless the reader also has a copy of the precise book that you used when you were encoding your message. To make things secure, you do not want to reuse any page of that book, hence the name "a one-time pad." This sounds immeasurably secure, except that it has one fault which is that the book itself is not random. It is possible to determine patterns of speech within a book that one has used to encode the message. To obtain a truly effective one-time pad, the book should be composed of truly random numbers.

Intelligence organizations generate these truly random numbers through measurement of inherently random physical processes such as radioactive decay and actual electronic signal noise. The randomness of this noise so completely overwhelms the information content of the message as to make it essentially undecipherable. This randomness, this statistical lack of any pattern within a data set, is referred to as Shannon entropy and is analogous to the mechanical entropy of physical systems. The revelation is that systems with greater entropy contain more information.

In this chapter we learned about the quantum of information, the bit. We saw that information content of a message is related to the number of possible messages and also to the similarity of the elements of a message to each other. We observed that highly intricate structures produced by repetition of a simple rule contain little information and that noise, counterintuitively, represents great information density. We learned to compare information between datasets not only by their size but by their compressibility. We can now revisit the genome with our new tools and see whether we can fit an entire brain into 46 strands of DNA.

Chapter 5: Nondeterminism

In the early 1980s, at the height of the Cold War, the state legislatures of California and Arkansas and ultimately the House of Representatives of the United States demonstrated consummate bravery by taking on the ultimate enemy: Satan and his avenue to the innocent minds of American youth, Led Zeppelin. The 1983 California bill was meant to address recordings that might "manipulate our behavior without our knowledge or consent and turn us into disciples of the Antichrist". This was a very serious matter indeed, perhaps the most serious, and Arkansas unanimously approved a similar bill. Of concern for those concerned, the young governor of the state, Bill Clinton, vetoed the bill and the State House was unable to overcome the veto. A similar bill was introduced by Republican representative Bob Dornan of California to the United States House of Representatives. But in what must have seemed a victory for the devil himself, the bill never made it out of subcommittee.

These legislators had become concerned that certain record albums, when played backwards, contained satanic messages. Songs that contain hidden messages when played backwards were said to exhibit backward masking, or simply backmasking. The most famous example is Led Zeppelin's Stairway to Heaven and its line, already abstruse, "If there's a bustle in your hedgerow, don't be alarmed now" that, when played backwards said, the concerned claimed, "Here's to my sweet Satan." California, true to form, sought to require the following label on records "Warning: This record contains backward masking which may be perceptible at a subliminal level when the record is played forward".

In the end, Satan was defeated not by legislation but by the invention of the compact disc player which did not allow for backwards play.

Questions of Jimmy Page's lucidity aside, is it possible to create lyrics that convey a message in either direction? Would such a feature so constrain the sort of message that could be produced that it would compromise the amount of information that the message would convey? Put another way, could we get twice as much information from our genome if we read it forward and backward?

Words that can be read forward and backward with the same result are, of course, called palindromes. We are absolutely not talking about palindromic DNA because for these, amusing as they might be, the reverse direction would carry identical information to the forward direction. The linguistic term for what we seek is called a semordnilap, a clever play on the word palindrome. Examples of palindromes are words like racecar and level that carry the same information content in either direction. Examples of semordnilaps include "desserts", which read backwards is "stressed", and "diaper", which read backwards is "repaid". Are there any genes that are semordnilaps?

To answer that question, we need to look at an organism with a highly compacted genome. In contrast to the 2% encoding portion versus 98% non-encoding portion of our genome, yeast, specifically Saccharomyces cerevisiae, has a genome in which 85% is encoding and only 15% is non-encoding. This focus on minimizing the size and therefore maximizing the efficiency of the yeast genome must be the outcome of some adaptive pressure. The result has been the development of a genome that is ultra-compressed.

Let's learn a bit of jargon. Nucleotide base pairs, the byte-equivalent of the genome, are read in groups of three, referred to as codons. There are enough permutations of three nucleotides, each of which can be one of four possibilities, to provide unique codes for each of the 20 amino acids from which our proteins are composed. Just like human language, these codons are read in a particular direction. The direction in which they are read is referred to by geneticists as the *sense* direction. The opposite direction, in which backmasking might occur, is referred to as the *antisense* direction. Mutations or other disruptions that result in undecipherable combinations of nucleotides are referred to as *nonsense* and errors or omissions that caused codons to be read as encoding the wrong amino acid are referred to as *missense*.

Saccharomyces cerevisiae, like all organisms, interprets its genome in the sense direction, but, amazingly, also does some of its decoding in the antisense direction. This means that there are some parts of the genome that are read both in the forward direction and in the backward direction and contain different information depending upon in which direction they are read. However, the degree of back coding present in this yeast is very limited and does not include any actual genes. The information that is gathered through antisense reading plays a role in what we called conditional statements or regulation of gene expression rather than encoding the creation of a novel protein.

The result is that some additional information in the yeast genome is encoded in the antisense direction, but the actual amount of information is very, very small. Even if such back coding existed in the human genome, the amount of additional information it would represent would be far too trivial to help us in our construction of the complex synaptic network of the brain.

In our last chapter, we explored the information density of noise. We understood that this noise was not always helpful in conveying the information we wanted to convey, but we did recognize that it is a rich source of information nonetheless. Might we introduce some degree of randomness into the genome itself in order to garner more information? What would this randomness look like?

Let's revisit the imperfect analogy of the genome as a book. How could we introduce randomness into a book? One thing that we might consider doing is randomizing the order of sentences. If we completely randomize everything, we are likely to wind up with unintelligible text. However, what if this randomization were limited? What if the mixing of sentences were constrained but still random in nature? This sort of shuffling occurs in organisms and the portions of the genome that are shuffled are referred to as transposable elements or, colloquially, as jumping genes. This is a bit of a misnomer because it is not an entire gene encoding for a protein that is typically shuffled, but rather a shorter bit of genetic code. Barbara McClintock won the Nobel Prize in Physiology or Medicine in 1983 for her discovery of transposable elements.

While shuffling sentences, or more accurately parts of sentences, within an essay will produce novel combinations, very few of these new combinations will make sense. Some sentences will be turned into nonsense and, if enough shuffling takes place, the essay itself will become incoherent. For this reason, cells have developed defense mechanisms to prevent the movement of these transposable elements. But like all defense mechanisms, the degree of prevention, the degree of mitigation, is less than complete. Thus, new information is introduced into the genome by the mechanism of transposition but, like the noise in our audio recording, this new information is generally not of benefit to the organism. Transposable elements are implicated in disease like hemophilia and cancer and likely play a role in the introduction of mutations that are the stuff on which evolution is built. But at the level of an individual organism, at the level of neurodevelopment, transposable elements are not the answer for which we are looking.

Have we entered a genetic niche in our discussion of what have become called transposons? Is this something so obscure as to not really matter? Far from it! Although only 2% of our genome constitutes protein-encoding genes, DNA transposons constitute 3% of the genome, or 1½ times as much of our genome as genes themselves! Other transposable elements represent more than 40% of our genome and some of these are remnants of retroviruses.

A retrovirus, like all viruses, originates outside of the organism or cell that it infects. Unlike other viruses, retroviruses are translated from RNA into DNA, this being opposite to the pathway of typical protein construction, and this newly translated DNA is incorporated into the organism's genome itself. Some of these retroviruses produce frank pathology, like HIV. Other retroviruses seem to have lost their pathological potential and reside as fragments in our own genome. As parts of the genome, these retroviral fragments are passed on to future generations and it is difficult to say which of our ancestors in our evolutionary tree acquired the original retrovirus. These endogenous retroviruses may represent as much as half of our genome, and certainly represent many, many times the amount of our genome that is devoted to actual encoding of proteins.

Like ambitious reporters competing for column space in a newspaper, the parties competing for space in the genome include not only transposable elements, genetic expression control sequences, and endogenous retroviruses, but also of aggressive genes as well. These *selfish genes*, a term coined by Richard Dawkins, or parasitic DNA, enhance their own replication within the genome and in some cases interfere with expression of less aggressive genes. Richard Dawkins in "The Selfish Gene" advocates for the idea of the gene, and not the organism, as the unit of evolution. He proposes that competition between genes is the real basis of evolution and that organisms are merely vessels for the genes.

But you and I can propose an alternative view, having digested the perspective of Benoit Mandelbrot. Evolution is itself fractal: it occurs on many levels and at many scales. Selection and evolution can occur at the level of the gene, at the level of the organism, or even at the level of the hive and one is not exclusive of the others. Evolution is like our snowflake that describes a similar pattern over a wide range of scales.

We've seen in this context that although adding a random element provides a new information source, this information has not furthered our project of producing a brain. Should we discard randomness as unhelpful noise as was done in the time before Claude Shannon?

Let's examine an instance in which randomness provides an unequivocal value. In the early days of artificial intelligence, two important schools competed: advocates of symbolic AI, represented by Marvin Minsky, who asserted that intelligence arose from the manipulation of abstract symbols representing concepts and relationships, and advocates of neural networks, represented by Geoffrey Hinton, who saw reinforceable connections as the key to AI. Minsky's criticisms of the connectionist school were centered on the fact that the early neural networks performed poorly and were not robust to changes in the sorts of problems with which they were presented. The algorithms worked in only certain circumstances but otherwise demonstrated themselves to be fragile. In part, because of Minsky's criticism, interest in and funding for neural networks greatly diminished in what came to be known as the AI winter.

Since we are accustomed to image recognition software that is based on algorithms derived from these neural networks, we have to ask, what changed? Why do neural networks perform so nicely now? The key to this improved performance was the introduction of random elements into the neural network algorithm. I am going to give an analogy to the classic problem of gradient descent but, before I do so, I want to ask the forgiveness of my many AI colleagues for the gross simplification I am about to construct.

Let's imagine ourselves as surveyors who wish to plan a hiking path from the peak of a mountain to its base. This path from a highest point to a lowest point is analogous to many of the mathematical challenges we encounter when developing AI algorithms. Let's further complicate things by imagining that we are doing this in complete darkness and with no knowledge of the topography of the landscape.

Our strategy involves placing a hiker at the peak of the mountain and asking him to find his way down. The route he plots will be used to design a hiking path for visitors to the mountain. What strategy should the hiker take in order to move from the top of the mountain to the bottom? This sounds like such an easy question as to pose no real challenge. The hiker should simply walk in the direction that he recognizes as down. Problem solved, right?

Well, not so quickly. Real mountains do not have continuous slopes from the top to the bottom but rather describe an irregular shape that incorporates many crevasses and excavations. How might this prove problematic to our strategy? Let us say that the hiker's stride is 1 meter, that is to say that when the hiker moves, he plants each foot 1 meter in front of the other foot. The hiker is instructed to go in the direction that is down. Keep in mind, this is in complete darkness and the hiker cannot see where he is going. The hiker begins making good progress, moving downward from the peak and then, without knowing he is doing so, walks into a shallow crater of three meters diameter. He continues his descent to the bottom of the crater but then, regardless of the direction in which he moves, he finds himself to be ascending as he moves up the side of the crater. He therefore concludes that he has reached the bottom of the mountain and that his job is done. Wherein lies his mistake? It is that he has not reached the absolute nadir of the landscape, but rather that he has found himself in a little localized dip. In computer science, we call this a local minimum. By employing the strategy of always walking down, the hiker has no knowledge that the base of the mountain lies far below. This problem has arisen because of the relation of the length of his stride to the width of the crater.

Certainly, the hiker would not have gotten stuck in this local minimum had his stride been longer or perhaps shorter than 1 meter. Might we have had more success by choosing a hiker with a 2 meter stride? Probably not. It's true that this new hiker might not have gotten stuck in the particular 3 meter crater, but he is likely to have found his way into a different sized crater that posed the same problem, and convinced him that the local minimum was in fact the base of the mountain.

The solution, counterintuitively, is to employ a hiker with a random stride. Since his stride length is always changing and in an unpredictable fashion, he will eventually find his way out of almost any crater in which he lands. He is far more likely to chart a path from the top of the mountain to the true bottom.

We have solved this gradient descent problem by introducing a random element. This element carried additional information that would have been absent had the stride length been fixed. A fixed algorithm, as we have seen, is what we call deterministic in the sense that we can predict the outcome with certainty. An algorithm that incorporates a random element is said to be *stochastic* in that we may understand the general outcome, but we are unable to predict with precision what the final product of the algorithm will be. We can make statistical predictions but stochastic algorithms are, by their nature, non-deterministic.

Although counterintuitive when compared to our usual understanding of the world, the introduction of stochastic elements not only help us solve problems but also encourage us to view the world in probabilistic terms.

On the other hand, we have now found ourselves in our own sort of local minimum. We seem to have pulled out all the stops with genetic data compression in order to produce the massive connectome of the brain. We have used compression and the introduction of conditional statements. We have looked at reading the genome both forward and backward and have incorporated stochastic elements in the form of transposons. And yet, we find ourselves little further down our path to building a brain than we had in Chapter 1. Let me make the case even more clear. What if the genome were amenable to perfect compression? What if every single base pair represented information about a synapse? What if the entire 770 megabytes of genetic information were devoted to brain architecture? Even in this physically impossible hypothetical, we would still be nowhere near an instruction set to build a functioning brain. Does our story end here or have we found ourselves in a little crater on a hillside that leads much, much further down? Perhaps it is time to randomize our own stride and we will do so by turning not to Markram or Minsky or Mandelbrot, but to the weather and to chaos itself. And this is where we will pick up the story in the next chapter.

In this chapter we took great pains to extract to compress as much information as possible into the genome. We talked about back coding, that is to say, antisense reading of parts of the genome, and of introducing randomness as an additional information source. We met Marvin Minsky, who, though an early critic of neural networks, is a giant in the field of AI. But we saw, for all our efforts, that interpreting the genome as an instruction set, even were it maximally compressed, is inadequate by many orders of magnitude to produce the level of complexity represented by the human brain. We decided that a new approach is needed, one that will incorporate stochastic elements, but one that will take a different view of how information is employed to build us.

Chapter 6: Utter Chaos

We may regard the present state of the universe as the effect of its past and the cause of its future. An intellect which at a certain moment would know all forces that set nature in motion, and all positions of all items of which nature is composed, if this intellect were also vast enough to submit these data to analysis, it would embrace in a single formula the movements of the greatest bodies of the universe and those of the tiniest atom; for such an intellect nothing would be uncertain and the future just like the past could be present before its eyes.

These words were penned by Pierre Simon Laplace in 1795, although he had expressed similar ideas as early as 1793. The idea of an intellect so vast as to perform these calculations became known subsequently as Laplace's superman or, more commonly, as Laplace's demon, although Laplace himself never used this term. This enthusiastic expression that followed Newton's development of the laws of classical physics a century before, we now know to be flawed. The reason that immediately comes to mind is the evolution of quantum uncertainty by the likes of Heisenberg and his contemporaries.

But the failure of such a demon to see the future is more nuanced and interesting than just the negation of the idea of determinism by the introduction of quantum uncertainty. Let's imagine a universe in which everything were, indeed, certain, in which there were no quantum foam or Heisenberg uncertainty principle. A universe in which the path of each particle were absolutely predictable. We would call this a deterministic universe since, with knowledge of the applicable rules, the future location and velocity of every particle could be known.

In such a universe, Laplace's demon would still fail. One reason for this failure is immediately apparent if we were to seek to design or grow or in any way develop such an intellect. If the demon were made exclusively of atoms that performed calculation, what has been named by Norman Margolus and Tommaso Toffoli as *computronium*, and if this demon had absolutely no substance that did not perform computation, how much computronium would be required? Atoms and subatomic particles each have a number of properties like spin and charge. But let's give our hypothetical universe the benefit of an even more implausible simplification. Let's pretend that each particle of this universe has only one property that had to be computed and let's further advantage our demon by saying that each of its own atoms could perform a calculation of this property.

In that way, the demon would only need one particle for every particle in the universe in order to see the future with certainty. Do you see where I'm going here? If such a perfectly efficient intelligence were to exist, that intelligence would have to be, at a minimum, the same size as the entire universe. Also, since electromagnetic radiation, light and its ilk, necessarily move at light speed, the demon could not calculate the universe any faster than the universe itself moves. At perfect efficiency, the demon would take one second to see one second into the future or one hour to see one hour into the future. The demon would be useless as a predictor. In fact, the demon would *be* the universe.

But there is a point of failure that has nothing to do with size or computational ability and it is a failure that is highly relevant to our project of constructing a brain. It is that, even in the event that the universe were entirely deterministic and the universe contained no randomness at all, and even if calculation were possible, there would remain no formula to predict the future.

So far in this book, we have contrasted determinism with randomness. A deterministic system is one that follows defined rules without deviation, and a random system has at least some elements of chance. But does deterministic always mean predictable? Can a deterministic system produce outcomes that are chaotic without the injection of any random element?

To answer this question, we are going to look at some classic examples of deterministic systems that produce chaotic results. The first of these is a sort of pendulum. Surely nothing could be more predictable than a pendulum, right? Let's look at two pendulums. If these pendulums are examined side by side, each will remain completely predictable. However, what if we stack these pendulums? What if we take a large pendulum consisting of an arm and a weight and suspend another arm and weight from it? This double pendulum would consist of a swinging arm and weight that is suspended from the weight of a larger pendulum above. In fact, the relative sizes of the two pendulums don't matter for this example.

We are now going to pull back the arm of the pendulum. Which arm? It doesn't really matter. You can pull both back or just one of them and then release it. The pendulum will swing as pendulums do, but because the top pendulum, let's call it pendulum #1, is also being tugged by the pendulum it itself is suspending, which we will call pendulum #2. Its path through the air, its velocity, is not uniform. It moves in a jerky fashion. This jerky fashion then affects the way that pendulum #2 swings. The motion of the suspended, lowest weight, is chaotic and is unpredictable. That's really an incredible claim, that a system so simple as two stacked pendulums would be unpredictable even with all of our modern mathematics.

Could this unpredictability arise from small random perturbations in the air or changes in temperature between the joints of the pendulum resulting from friction? Could there be some hidden random element? Amazingly, if we build a simulation on a computer of a double pendulum so that we eliminate all of these extraneous environmental variables, the path of the weight on pendulum #2 remains chaotic and unpredictable. The system, particularly the computer simulation but to a large extent the demonstration in the real world as well, is entirely deterministic and the applicable rules of motion are very well understood and have been for a great deal of time. But the outcome remains chaotic and unpredictable.

Another example of a simple chaotic system is what has become known as the three-body problem. Aerospace engineers have been able to launch satellites, send a capsule containing human beings from a moving and rotating planet (ours) to a moving and rotating celestial body (the moon) and have even been able to design spacecraft like Voyager that has passed through the gravitational fields of many planets in order to follow a precisely planned path. Given a star and a planet, astrophysicists can describe with certainty the path of the planet around this star even accounting for the perturbations introduced by general relativity.

But do you know what these scientists can't do? If we introduce a third celestial body into the equation, another star or another planet, the boffins cannot predict indefinitely the future locations of the three celestial bodies with respect to each other. Astrophysicists can make very good predictions for very long stretches of time for the locations of the planets in our solar system, but there is an amount of time beyond which such calculation is impossible. This unpredictability of systems including three or more celestial bodies is the premise that underlies Liu Cixin's incredible trilogy, The Three Body Problem.

The quality that the double pendulum and that the three celestial bodies have in common is *aperiodicity*. This means that, no matter how long we observe the system, it never repeats itself exactly. And this is true despite the fact that there is nothing random about the system.

The failure to predict the position of the Earth over astronomical time is interesting but is of little relevance in our lives unless we are preparing a dissertation. However, this aperiodicity is highly relevant in the prediction of the weather.

Today's protagonist was a meteorologist and mathematician, a potent combination. Edward Lorenz had only recently completed his master's degree in mathematics at Harvard University when the US entered World War II. Lorenz served in the United States Army Air Corps as a weather forecaster. Now, it does not take a degree from Harvard to recognize that weather prediction is, to put it generously, imprecise. The problem is that there is a lot going on in the atmosphere. Atmospheric circulation is a three-dimensional affair and its analysis requires the assessment of a great number of variables. Lorenz sought to simplify this and created a model consisting of three, relatively simple, partial differential equations representing atmospheric convection.

In order to understand what came next, I need to introduce the idea of phase space. We understand the Cartesian coordinates of X, Y and Z or up/down, left/right, forward/back to distinguish the location of objects around us. Phase space is not a complicated concept, but it is one that a lot of us are unused to. So, to the extent that I can, I'm going to simplify our discussion.

Let's talk in only two dimensions and forget about elevation. We can describe the location of something based on east/west and north/south axes. In other words, let's draw a map. My office is on Madison Avenue near 32nd Street, and on nice days, I will ride my bike from the West Side to my office. We can imagine a map in which my office is designated by a point between 32nd and 33rd Streets on Madison Avenue. Let's add in the element of time and imagine that I am on my bicycle at 32nd Street and 6th Avenue, that is to say a couple blocks west and a little bit south of my office.

At time 0, probably 7:15 AM, my location can be designated by a point at 6th Avenue and 32nd Street. At 7:17 AM, my location will be between 6th and 5th Avenues on 32nd Street, a bit closer to my office. At 7:20 AM, I'm likely to be at the intersection of 32nd Street and Madison Avenue and getting ready to turn left to head the half block north to my office. By 7:21, I have gotten off my bike and am walking into the lobby of my building.

Now, let's connect the dots. We would see that we have formed a path that I have taken to get to my office. All well and good. GPS does exactly this for us when it shows us the direction that we will take, which is simply the points from where we are to where we are going projected over time.

This is a map or a graph but it is not phase space. To achieve that, let's substitute one of the axes for speed and the other axis for the distance to my office at any particular point in time. What would this graph look like? At 7:15 AM, my distance to the office is 500 m and my speed is 20 kph. At 7:20 AM when I am only half a block from my office, my distance to the office is 50 m and I'm still trucking along at 20 kph. As I approach my office lobby and my distance shrinks to a couple meters, I am going to slow down to perhaps 3 kph and as the distance to my office reaches 0 my speed will reach 0 as well.

The graph of my speed versus distance to the office will describe a straight horizontal line that, near the end, curves downward. This is a graph of phase space in which each point represents the values of distance and speed. The entire graph, including all possible speeds up to 40 kilometers per hour (because I have certainly never pedaled any faster than that) and distance up to 500 meters encompasses all of the values that I could possibly have on this ride. The line we have drawn represents my values for distance and speed for my short bike ride from 6th Avenue.

Phase space is simply the part of the graph that contains all possible values for a given system. Those values might be, as in the case I have given you, speed and distance. But the values could just as well have been temperature and noise or three different values for a three-dimensional graph. In fact, there are no constraints on the number of axes, although it is impossible for us to visualize any graph that would have more than three. But we could certainly follow 5 or 10 or any arbitrary number of parameters in a system and the envelope that contains all possible parameters would represent that system's phase space.

Lorenz plotted out the path through phase space of the state of the atmosphere as it evolved over time and, in doing so, made a number of observations that are relevant to our project. But to explain this relevance, we need to understand the idea of periodicity.

Let's make one more phase space graph and this one is of a fun system. Let's say that we attach a baseball to a rubber band. The axes of the phase space we are going to measure are the distance of the ball from my hand and the velocity of the ball as it pulls on the rubber band I hold in my hand. As the ball stretches the rubber band, its velocity decreases because of the resistance of the rubber band. When the ball is at its maximal distance from my hand, velocity has decreased to 0. Then, as the rubber band tugs on the ball, its velocity increases but in the opposite direction, that is to say, back up toward my hand. As the rubber band contracts, it generates less and less force on the ball and gravity starts to play a larger role eventually stopping the ball in space at a point at which distance to my hand is at its minimum and velocity has again dropped to 0.

The plot of this graph in phase space is pretty interesting. Let's say the Y axis represents distance from my hand and it varies between 1cm and 50 centimeters. Let's say the X axis represents velocity of the ball toward the ground. In its descent, the ball will have a positive velocity because it is moving toward the ground and in its ascent will have a negative velocity because it is heading away from the ground. When we plot this out in the phase space we have described, we see the plot is in the shape of a circle because at the greatest and least distance from my hand the X value, that is to say the velocity, is 0 and midway through its drop or ascent the value along the X axis is maximal but has positive values while the ball is dropping and negative values while the ball is ascending. Putting this all together, we get a circle. The circle describes the motion and velocity of the ball over time and as the ball yoyos, these values go around and around that circle. Each time the ball reaches its maximal distance, it describes one circuit or, in our language, one period. The motion of the ball is periodic because it will forever move along the circle we have described.

Except, of course, this is completely untrue. You and I both know that the ball will eventually oscillate less and less and less until it comes to rest about halfway down. This is because of friction both with the air and within the rubber band itself and loss of energy over time. For sticklers, I understand that it is conversion of energy from velocity to heat, but let's just go with loss for now. What would the real path of the ball through phase space look like? It will be a spiral gradually spiraling in until the ball's distance from my hand is midway and velocity has dropped to 0.

This is fascinating, although it may not be obvious why. The phase space spiral described by the ball in the real world never crosses itself. There is no intersection in that spiral. What does this mean? It means that the ball is never at precisely the same position and velocity. This is an aperiodic system because the system is not cyclical.

Lorenz found that his measurement of atmospheric circulation was aperiodic. More than that, he found that, far from intersecting, the path through phase space grew further and further apart. As a meteorologist, this was a real problem because it suggested that no pattern existed that would allow him to predict the weather.

An awful lot seemed to depend on the starting values he chose for atmospheric flow. Similar values would produce similar results in his prediction of atmospheric behavior but only for a short time. After that, the model's prediction would be a bit different and then very different even if the initial values he chose were almost identical. He referred to this property as "sensitivity to initial conditions" and this realization proved profound. Since weather prediction is ultimately based upon the measurement of current temperature and humidity and other physical factors, small differences in these measurements would result in substantially different predictions for the weather. Not initially, maybe the prediction for tomorrow's weather would be very similar regardless of small measurement errors, but longer range forecasting would yield radically different results arising from very small measurement errors. And it is from this observation that we get the term the butterfly effect. It is not, as it is popularly conceived, that the movement of a butterfly's wings is so amplified through the atmosphere that they can create a hurricane. Rather it is that the exquisite sensitivity to initial conditions that is the difference between predicting a hurricane and a clear day may be as small as a perturbation produced by a butterfly as it flaps its wings.

Lorenz made a second and even more interesting observation. He plotted the changes in atmospheric circulation in a three-dimensional phase space to look at the path that it took over time. This is akin to drawing out the route one has taken in a map, but it is important that the phase space Lorenz employed was three dimensional. I want to pause for a second and ask you, kind listener, what do you think this plot may have looked like? Lorenz had established that his results became unpredictable over a relatively short span of time. Would the path through the three-dimensional phase space also look chaotic? Might they just look like a jumble on a map?

Incredibly, the path that this chaotic system took formed a pattern, and that pattern was one of a bent figure 8. It is entirely coincidental that the pattern looks similar to a butterfly's wings because the term butterfly effect to describe changes in weather greatly predates Lorenz's work. We must ask what it means -- that a pattern existed at all. The pattern, like our spiral formed from the ball hanging from a rubber band, was aperiodic and nowhere did the lines of the pattern intersect. Remember, this was in three dimensions so that it is possible to form a figure 8 pattern without intersection in the same way that it is possible to have looping highway off ramps that do not collide with themselves.

Lorenz saw that although chaotic, the paths through phase space seemed to gravitate to this figure 8 pattern and he called this pattern an attractor. Readers versed in chaos theory may have heard the term *strange attractor* and the adjective strange refers to attractors that are fractal in nature, as was Lorenz's atmospheric attractor.

How can it be that a system can be both determinative and chaotic; and both chaotic and tending toward a pattern; and be tending toward a pattern but never repeating itself? These observations form the basis of what became known as chaos theory. Just because something is chaotic does not mean it cannot have tendencies.

If we return to the double pendulum and ask whether it too demonstrates an attractor, the answer is yes. I will not describe the shape of that attractor but invite you to Google "attractor for double pendulum".

This is immensely interesting, but what does it have to do with neurogenesis? Am I suggesting that chaotic systems influence the development of the brain? Actually, I'm attempting to make a more fundamental point than that. It has to do with how we describe systems. The fact that attractors do not intersect themselves means that describing them mathematically, or even in words, is extremely difficult. Since the systems are chaotic, it is not possible to predict in advance what the state of the system will be at any point in time even though the attractor shows us tendencies.

How can we compare systems? Can we come up with a formula that represents the attractor? The answer, for the most part, is no. And this is important for us as we shift our focus from objects to processes. Processes that evolve over time are generally described either statistically or by simulation. We will need to abandon the idea of precise prediction and begin to speak in terms of probabilities. These attractors represent phase states of increased probability. Over the next few chapters, we will change the way we view the complex systems of which the world is built from a focus on objects like transistors and neurons to concentrate instead on process and probability.

Perhaps you share my amazement that chaotic systems can form patterns. But what about genuinely random systems? Can a system be truly random and still have structure? To answer this question, we will need to go a great deal further back in time, many thousands of years, to the time and place of the pyramids, and this is where we will pick up next time.

In this chapter we discovered that some determinative systems can produce unpredictable and chaotic outcomes, even in the absence of any random elements. We saw that although chaotic, the systems can tend toward certain outcomes and we called these tendencies attractors. We recognized these attractors as zones of increased probability and understood that our language must change from objects and their properties to systems and probabilities.

Chapter 7: The Riddle of the Sands

Imhotep had an idea. As a chancellor, poet, physician, mathematician, astronomer, and architect to the pharaoh Djoser, ideas were nothing new to Imhotep. His writings on medicine were so well regarded that he was, 2000 years later, deified and conflated with Asclepius, the Greek god of healing. The Roman emperors Tiberius and Claudius had inscriptions to Imhotep carved into their temples nearly 3000 years after his death. Historical time is not the subject of this book, but I can't help pointing out that less time has passed between now and the reign of Emperor Claudius than had passed between Claudius and Imhotep! Suffice it to say, ideas were not new to Imhotep.

Since before the time of the first dynasty of the Old Kingdom of Egypt, prominent people including Pharaohs were buried in mastabas. A mastaba is a single floor multi-room mausoleum in which one room contained the sarcophagus, the coffin, and the mummified pharaoh. The mastaba was a sort of a mortuarial ranch house. Imhotep wanted to design something special for his Pharaoh, Djoser, and this idea was to build – up. Rather than a single-story tomb, Imhotep would build a multilevel affair by stacking mastabas of decreasing size to create a sort of mastaba wedding cake. Seven mastabas high, this structure, visible to this day, is known as the stepped pyramid of Djoser.

About 70 years later, another Pharaoh, Sneferu, had an even more audacious idea. His tomb would not be stepped, but would have smooth sides. It would be a genuine pyramid. Construction began in Meidum, a three-hour trek south of the stepped pyramid of Djoser and was well underway when the entire structure collapsed. The site today is a pile of debris with the remnants of the original interior at its center.

Although no contemporaneous description of the collapse remains, inference may be drawn from the unusual design of a neighboring pyramid built at about the same time. The bent pyramid, as it is now known, has a comparatively steep slope that dramatically flattens part way up. Specifically, the angle of the side of the bottom half of the pyramid is 54 degrees to the horizontal when the designers seem suddenly to have changed their minds about the angle of the walls and decreased the slope to 43 degrees, giving it a profile similar to a barn. What prompted this decision? Could it have been the experience of having built at Meidum a pyramid with a steep rake only to see it collapse? One thing is certain, no pyramid would ever again be built with a slope as steep as 54 degrees.

5,000 years after the collapse of the pyramid of Meidum, a set of researchers were also working on sand. Per Bak of the Brookhaven National Laboratory, Kurt Wiesenfeld, and Chao Tang, then of the Brookhaven National Laboratory, were dropping individual grains of sand onto a little table. The table consisted of a metal disc standing on a metal column. Their apparatus was arranged to drop a single grain of sand onto the center of the disc, one 600 microgram grain at a time, and to repeat.

As the sand pile spread to the edge of the metal disc, some grains would fall off and accumulate on a skirt that was isolated from a scale measuring the mass of the sand pile. At about 35,000 grains for a 4-

centimeter disc, sand was falling off the edge at about the same rate as their apparatus was adding it, and the mass of the sandpile stabilized.

This sounds like the apex of tedium and inapplicability (has your life changed with your new knowledge that 35,000 grains of sand is the capacity of a 4 cm disc?). But the insight of the researchers was to look at how the grains of sand fell off the edge.

Early on in its growth, almost all of the grains of sand in the pile stayed right where they were dropped. Occasionally, a grain would knock over another grain of sand and a mini avalanche would displace a few grains of the pile. As the pile grew taller and its sides steeper, larger avalanches would occur in which many grains of sand would fall. Once, however, the system stabilized -- with as many grains of sand falling off per unit of time on average as grains that were added to the pile -- a particular steepness, that is to say slope, was reached.

Coincidentally, this slope for their sandpile turned out to be 51° and steeper slopes could not be achieved for any duration of time without an avalanche reducing the height and slope of the pile. The system achieved an equilibrium, but one at a state of criticality such that exceeding this threshold would cause an unpredictable, but eventually certain, catastrophe in the form of an avalanche.

Bak, Tang, and Wiesenfeld observed a number of interesting properties to what became known as the BTW or Bak–Tang–

Wiesenfeld model, now more commonly known as the Abelian sandpile model (ASM).

The shape of this pile, dare I say, rounded pyramid, was consistent when the experiment was repeated and had a characteristic slope. The system achieved criticality without any outside influence or design. The critical state was robust in the sense that whenever a large avalanche reduced the slope or an unusually large number of balancing grains increased the slope, the system always returned to its stable, critical state.

This *self-organizing criticality*, as the researchers called it, seemed to arise on its own without any internal instruction set. And, more importantly for our mission, arose not as a result of any single grain of sand but rather from the process of interaction between grains. In these chapters, this is the first time that we are witnessing an example of a system that is greater than the sum of its parts. But I'm getting ahead of myself.

Once criticality was achieved, avalanches would still occur but they were of varying sizes. Small avalanches occurred frequently but displaced few grains of sand while large avalanches that changed a great portion of the sandpile were less frequent. Incredibly, scientists recognized this same frequency/magnitude relationship in earthquakes, the distribution of clouds and of galaxies, and of vortices in turbulent fluids. The relationship is said to follow a power law and is, you guessed it, fractal.

This is cool stuff, but what is its relevance to us? It is to say that the information for construction of these systems was intrinsic to the system itself. Can biological systems self-organize?

Many single-celled organisms can migrate *to* a source of food or *away* from a noxious stimulus. This *chemotaxis* can produce quite beautiful patterns. Some of my early work dealt with diffusion limited aggregation, a fancy term for what's really a very simple process. Let's imagine a game, a rather bizarre game, that is played with a group of people who sit on a smooth floor in a circle facing each other. Each participant in the game has a tennis ball with patches of Velcro on its surface.

One by one, the "players" roll a ball toward the center of their circle. By chance, a ball may come into contact with another ball and, through the magic of Velcro, stick to it. As more balls are rolled and more stick to the growing mass in the center, the mass takes on a shape.

What do you think this shape will be? Will it be a disc of velcro'ed balls? Will it be an irregular blob? Actually, something pretty interesting happens. Balls at the periphery of the mass act as relative barriers and as balls stick to these barriers, they begin to form lines that branch. As the mass expands, it takes on a shape that looks vascular or arborizing or dendritic.

We ask the same question here we did of the sandpile: where did the instruction for this pattern come from? It is a manifestation not of any individual ball, but rather an expression that has arisen from the interactions between elements of the system. It is, again, a whole larger than the sum of its parts. And it required no genome to form. The pattern is genuinely self-organizing.

As cells migrate toward a stimulus, they too can form patterns. The simplest, and most important, of these is determination of orientation. Think about the morula, the initial bundle of cells that forms after egg and sperm unite. The morula is fairly spherical, and yet one end must wind up being head and the other end tail. How do the cells of the morula know where they are, which end they will become?

A theoretical framework was proposed by, of all people, Alan Turing in 1952. His idea was that a signal in the form of a chemical might be produced at one end of a developing mass of cells. Cells within that mass could then sense, by virtue of detection of the concentration of this chemical, where they were with respect to its source. Turing called these hypothetical biological signals morphogens and it was only decades later that such morphogens were indeed discovered in that organism that is the darling of all geneticists, the fruit fly. We can think of the concentration gradient like the magnetic poles of the earth that give the compass a direction of North and South. We actually have experience in our own lives with concentration gradients every time we smell smoke. We are in one room of the house, and we smell smoke and we run around to the other rooms to see which smells most strongly of smoke with the intention of finding its source. What we are doing is following a concentration gradient with our nose.

This rudimentary morphogen orientation signal is much more highly developed in mammalian neurogenesis. At the tip of the growing nerve cell is a *growth cone* that we can think of as akin to our nose sniffing out smoke. The chemicals that the growth cone detects may act over short or long ranges and may be attractive or repulsive. These netrins and semaphorins establish concentration gradients and other similar signals can even be found on the surface of neighboring cells.

It may not be difficult to imagine a nerve cell following the path of a concentration gradient and winding up at the correct location, but I want to add some scale to this to show you how really incredible it is. Nerve cells may be only 60 microns wide but may stretch to over 1 meter in length. If we scale this up to the size of a motor vehicle, a neuron would be as wide as a car but would be over 30 kilometers long. If all the nerve is doing in its development is following a concentration gradient, that would be akin to following the smell of smoke over many tens of kilometers. This would be a bit much to ask of even the animal with the keenest sense of smell. Indeed, the nerve has some help.

Not only are attractive gradients present that direct the nerve over long distances but repulsive ones are present as well. Also, in its propagation, the nerve does not grow directly from point A to point B, but rather passes through intermediate waypoints in its path to its final destination. A single nerve is often not alone in its path and can also receive signals from neighboring cells. This sounds like a complicated process but it is entirely akin to our taking a road trip.

If we were traveling from Cleveland to Boston by car, we do not encounter exit signs for Boston immediately upon departing Cleveland. Rather, we see signs for closer locations that are along the way. Perhaps we see a sign for Erie, Pennsylvania and then for

Buffalo and then Albany and then finally signs for Boston. These intermediate waypoints help to guide us in our path and there are clusters of cells past which the neuron will grow that serve the same role as the intermediate signposts for these waypoints. These shorter range directions are governed by a different set of chemicals that are short range attractants. There are also short range repellents. Perhaps if we see a sign for a landfill or a garbage dump, we will be less likely to pull off the road there. These would be the equivalent of short range repellents. Similarly, if our road atlas or GPS shows us the city of Chernobyl or Semipalatinsk, this may act as a long range repellent as we would want to steer well clear of these contaminated sites.

The growth of the nerve cell is also aided by the presence of surface chemicals on neighboring cells that can act as attractants or as repellents. Imagine these to be the bumper stickers on the cars passing by us on the road. We might be more inclined to follow a car with the bumper sticker that says "knowledge through science" than a car with a bumper sticker proclaiming "death to intellectuals". Eventually, traffic may get so heavy along the road that we are physically constrained in our progress and follow a bundle of cars moving along our path. Neurons, in their growth, can also become parts of bundles with physical direction toward their goal. Perhaps some of the traffic will divert at Albany to go to New York City and we may leave the bundle to continue on our way to Boston following the attractant of the signs indicating the direction to Boston.

This is highly significant for our quest to build a complicated structure out of a minimum of instructions. In order to navigate from Cleveland to Boston, we did not need to first develop a road map. We merely had to follow signs. We will still reach our planned destination if the road veers a bit south or a bit north as long as we follow the attractants and avoid the repellents. This is an economical use of information. What does it sacrifice? Since the path of the neuron is not planned as one might do in a blueprint, that is to say it is not deterministic, there will be some variability even if the nerve starts and ends at the appropriate place. By incorporating a degree of uncertainty in the process, we can get what we need with far less information than if we were to have to encode an actual path. Here, as with our example of the hiker and gradient descent, including a stochastic element, that is to say a probabilistic one rather than a deterministic one, we are able to achieve our goal with much greater efficiency.

The other distinguishing characteristic of this system is that it relies on the interactions between multiple cells, those forming waypoints and those neighbors with signals on their surfaces. We see again that our investigation reveals a process rather than an object. Finally, by incorporating concentration gradients, we are achieving some degree of self-assembly and getting some structure and achieving some degree of architecture with no instruction at all.

But how far can we take this? Can we self-assemble all the way up to a functioning brain? And if we can do so, can we describe how it is done so that we can begin with our genome and end with our organ of thought? This is a heavy question and to answer it we will have to look at the meaning of life itself and that is what we will do in our next chapter.

As with our road trip from Cleveland to Boston, we have covered a great deal of ground today. We have seen that biological and even non-biological objects, when allowed to interact with each other, can produce structures of remarkable intricacy and we called this process *self-assembly*. We began with systems that exhibited the seemingly contradictory properties of criticality and robustness and were impressed that all of this happened without external intervention or any set of instructions. We saw biological systems that incorporated self-assembly through the use of concentration gradients and exchanged some degree of variability for a large degree of information efficiency. Most significantly, we were introduced to systems in which the whole is more than the sum of their parts.

Chapter 8: The Meaning of Life

In 1859, at the age of 23, Mr. Bradley had lived in a number of towns throughout New England settling, briefly, in Springfield, Massachusetts where he finally found employment as a mechanical draftsman. Seeking more, Bradley went to Providence, Rhode Island, to study lithography, returning to Springfield to open that town's first lithography shop. Bradley found great success with his lithograph of the little-known presidential candidate Abraham Lincoln, a lithograph that sold quite well until Mr. Lincoln decided to grow a beard. At this point, those who had purchased Bradley's fine lithograph demanded refunds since the picture was no longer valid. Despondent, Bradley burned his remaining stock of the lithographs of the beardless Abraham Lincoln. What was a poor lithographer to do?

Having been introduced to European board games by a friend, Bradley set out to apply his prowess in lithography to this new medium. In 1860, Bradley, that is Milton Bradley, released his Checkered Game of Life, a game in which players would spin a teetotum, a sort of calibrated top, and advance through life's perils and triumphs, past prison, ruin, and suicide with the goal of achieving 100 points before demise. The game sold well and was followed by others catering to Americans' newly found leisure time.

There things remained for the next hundred years until 1960 when the game was modernized to reflect the current zeitgeist. Players progressed through the updated game getting married, having children, buying homeowners insurance, all presumably on a single income! The teetotum was replaced by a spinner and the game remains popular to this day.

Ten years later, John Horton Conway was working on a very different sort of game of life: one that did not require the two to six players of Milton Bradley's game. In fact, Conway described his Game of Life as a zero-player game.

John Conway was a mathematician interested in cellular automata. Cellular, in this context, means confined within a matrix of cells rather than being cellular in the biological sense. This matrix, in two dimensions, looks like a very large checkerboard. In fact, we can think of it as an infinitely large checkerboard in which the squares can either be empty and white or completely filled and black. In digital terms, each square can either have a value of zero or of one. The game progresses in turns in which each square of the matrix, each cell, either keeps its current value or changes its value from 0 to 1 or vice versa, depending upon the value of neighboring squares.

The idea of cellular automata predates Conway's work, having been invented, as were so many things, by John von Neumann in collaboration with Stanisław Ulam. Von Neumann has already appeared in this book. Ulam is best known for, along with Edward Teller, developing the hydrogen bomb.

Von Neumann had developed an interest in the replication of complex systems. In order to replicate itself, a complex system, for example a living organism, must have both the instructions for the replication process as well as the machinery to perform the replication. Since the system must contain all the information about itself as well as containing a replication mechanism, it sounds as if it should be impossible for a complex system to actually replicate. And yet, here we are.

Von Neumann envisioned machines that, emulating life, could replicate themselves. This line of thought led to the idea of von Neumann probes which are machines that would be launched into space and would replicate themselves with available material as well as machines that could alter their own instruction set to produce a sort of directed evolution.

Von Neumann's mathematical framework for developing ideas of replication and complex systems was cellular automata. He developed patterns and rule sets that could duplicate themselves and named these universal constructors. Incredibly, he did this all without the aid of computers.

The von Neumann machines, as they have come to be known, would appear to be rather threatening. Perhaps the idea of a space probe replicating itself is not something about which we have to worry, but self-replicating nanotechnology would certainly represent an existential threat. Having said this, the inventor of the term "gray goo" to describe a catastrophe of ecophagy has emphasized that the likelihood of such runaway replication is very small.

It seems to me that interest in, and fear of, nanotechnology has been replaced in popular media by interest in and fear of artificial intelligence.

John Conway's innovation in 1970 was to develop a set of four rules that would govern whether any particular cell changed its state and framed these rules as if they governed reproduction and sustenance of life. The rules are: 1) any live (that is to say filled) cell with fewer than two live neighbors dies, as if by underpopulation; 2) any live cell with two or three live neighbors lives on to the next generation; 3) any live cell with more than three live neighbors dies, as if by overpopulation; and 4) any dead cell with exactly three live neighbors becomes a live cell, as if by reproduction.

The player, more a sort of prime mover, arranges an initial pattern of cells and then the game proceeds, turn by turn, on its own without any further intervention by the player. Depending upon the initial pattern chosen, all of the cells may vanish and the entire system remains, as Conway would put it, lifeless. In other situations, the pattern will self-propagate and remain present, often in a cyclical form, for eternity.

However, in many cases, the pattern will develop over a large number of turns, either to collapse entirely or to become cyclical. Conway's Game of Life was demonstrated to be capable of universal computation, but, interesting as that is, I will not explore that topic any further.

We now have the technical language to describe and to understand Conway's Game of Life. It is a deterministic system since there are no stochastic or random elements. It is a system that can develop into a state that is cyclical, that is to say periodic, or a state that is chaotic and aperiodic. You will recall our discussions of deterministic aperiodic systems of chaos theory and the Lorenz attractor.

In experimenting with Conway's Game of Life, Stephen Wolfram, a polymath and creator of Mathematica, made a number of important observations, one of which is pertinent to our project of developing a brain. He found patterns that remained acyclical for more than 100 turns before either developing into cyclical patterns or vanishing entirely. Moreover, he found that no formula, no algorithm, could predict what the outcome of the initial pattern would be without going through each of the turns of the Game of Life. That is to say, there was no computation that was able to determine the outcome in any fewer steps than the game itself took. He referred to this property as computational irreducibility.

I want to highlight two things about computational irreducibility. The first is that we have already encountered it in our discussion of Laplace's demon back in Chapter 6. We said that even if the universe were deterministic, Laplace's demon would be useless for prediction because it would be unable to calculate the future of the universe any faster than one second per second. This is just another way of saying that the universe, in its detail, is computationally irreducible.

The other point I wish to make is to prevent any confusion between computational irreducibility, which is an important topic in complexity science, and irreducible complexity, which sounds similar but is a fallacy promulgated by creationists. Irreducible complexity claims that a structure as complicated as the eye could not have evolved because it functions only in its full development and does nothing when only partially evolved. The eye is the example generally cited despite the fact that an immense evolutionary record of eye development exists, demonstrating the fallacy of the idea of irreducible complexity. I will discuss this idea no more since I have no truck with it and bring it up only because it sounds similar to the phrase we are now using, computational irreducibility with which it has no overlap.

It is well and good for me to say that many biological processes are computationally irreducible, but an example will make the point clearer and this example comes from a fairy tale. In the story of Jack and the Beanstalk, Jack trades his cow for some magical seeds. He does not know what these seeds will produce when he trades for them. Should Jack have known? Indeed, if you were to get a seed from a plant nursery, a seed you did not recognize, is there any way you could know into what sort of plant it would grow? In fact, if you were an established botanist and came across a seed of shape and form and color you did not recognize, would you be able to, even by dissecting it, determine what would be the appearance of the adult plant?

The seed is the embodiment of the problem of computational irreducibility. We cannot know what the plant will be until the seed grows into it. If the process of growth to identifiability takes 30 days, there is no process, no algorithm that will allow identification in a shorter time. Perhaps, I hear you interject, we could do genetic testing on the seed to find its closest relatives. But if that were the case, the seed would not truly be novel to us. A truly novel seed cannot be accelerated in declaring itself.

This trivial fact is of profound importance for the field of computational biology, or theoretical biology. Many biological systems, many non-biological ones too, cannot be solved for but can only be simulated or predicted statistically. Even in the greatly simplified and entirely deterministic framework of Conway's Game of Life, the flourishing of an initial pattern cannot be solved for, only simulated.

And what of our project to understand neurogenesis? That fertilized egg that will ultimately grow to be a human, isn't that egg a seed? And does this mean that we have to abandon our project of getting from seed to brain? Well, not entirely. We may not be able to get from a specific genome to a specific brain, but we still have a great deal of road left to travel.

What does this mean to our project of moving from genome to brain? Have we, by recognizing that computational irreducibility is intrinsic to neurogenesis, moved the goalpost so far that our project is now meaningless? I think the answer is no because we must still start with the genome, with genetic expression, with genes. And there are no genes for complexity. There are no genes for systems. There are no genes for intelligence or language or height or any of these things. There are only genes for proteins and it is this journey from genetics to complexity that will begin our next chapter.

In this chapter, we learned of von Neumann's quest for self-replication and John Conway's zero player game of cellular automata. We saw with interest that some patterns can reproduce themselves and that it may be impossible to predict outcomes faster than systems themselves evolve. We understood that this must somehow lead back to genetics and began to wonder how to get from a protein to complexity.

Chapter 9: Agents of Change

This is a true story about machines. A large conglomerate had identified a reservoir of hydrocarbon fuel, but in order to put it to use, it had to be extracted and refined. The fuel itself consisted primarily of modified hydrocarbons and also of polymers. A polymer is a large molecule composed of repeating subunits and often serves some structural purpose. The conglomerate made machines to perform the refining, concentrating the hydrocarbons. This task was made more complicated by competition from other conglomerates that had identified the same hydrocarbon fuel source and had begun to manufacture their own machines to extract, concentrate, and refine the fuel. In a stroke of what seems, in retrospect, genius, the initial conglomerate modified its refining process to cause the polymers in the fuel to fuse to each other and turn the fuel into a thick gel that was largely impervious to the machines of the competitors. This strategy proved so successful that we continue to benefit from it even to this day.

The machines are now collectively known as rennet, the hydrocarbon fuel source as milk, and the coagulated and the fused polymers of casein are the first steps that transform the milk into cheese. The conglomerate in the story is the first human to make cheese. Rennet derives from the stomachs of mammals and while we think of rennet as primarily deriving from calves, the earliest domesticated milk producers were probably sheep and this first human making cheese may have come from northern Poland. Our earliest archaeological evidence of cheese production comes from Kuyavia, in northern Poland dated to about 5500 BCE and from the Dalmatian coast of

Croatia in 5200 BCE, but probably dates back an additional 2000 or 3000 years.

I tell this story in this way not because I wish to use machines as a metaphor for enzymes, but because enzymes actually are machines. Like all machines, the enzymes in rennet are composed of molecules and perform their task by moving and pivoting and making electromagnetic connections with a raw material in order to physically transform it.

Milk is composed of, amongst other things, a phosphoprotein called casein that surrounds globules of fat and keeps these "micelles" in suspension. The most important enzyme, i.e. machine, in rennet is a protein called chymosin or rennin, that cuts or cleaves the casein in a manner to foster fusion between casein molecules, creating a protein fat amalgam and separating this from the watery components we know as whey.

When we consume the cheese that is the final product that begins with forming the curd, other machines within us disassemble the machines we eat into pieces that we use to build machines of our own.

We ended Chapter 8 with the extraordinary claim that there are no genes for complexity or for intelligence or language or height. There are only genes for proteins. When we talk about genes for traits, there is a great deal of hand waving that goes on between the description of the gene and the manifestation of the trait. How do we get from a protein to the facility for language? Occasionally, there is a fairly direct connection as is the case with sickle cell disease in which a single base pair substitution in a gene leads to a protein that polymerizes into long chains that can distort the blood cell itself. In these single gene traits, it is often possible to draw a direct connection between the abnormal protein and the trait or pathology it causes. To use scientific terminology, the genes associated with a trait are called the genotype and the manifestation of the trait in physical form is called the phenotype. When Gregor Mendel performed his seminal work on pea plants, he discovered single gene traits that led directly to particular phenotypes.

Although such associations between genetic variations or mutations and their physical manifestations do exist, they represent only a small minority and some traits that appear monogenetic are instead the result of gene linkage rather than arising from a single gene. The great majority of traits are polygenetic and some involve environmental influences as well. Polygenetic traits are much harder to trace through lineages than monogenetic traits. This is both because multiple genes are involved, making probability more difficult to calculate, but also because there is no clear connection between the combination of genes and the trait or pathology they produce. A case in point is inflammatory bowel disease which, although having a strong genetic component, is not attributable to any single gene.

I'd like to reframe the problem of polygenetic traits in terms of machines as we did above with cheese. Each gene encodes for a protein, a chain of amino acids connected in the particular sequence governed by the genetic code. This string of amino acids is not itself a functioning machine, but becomes one by folding in on itself. In doing so, the protein exposes some regions that may fit lock-and-key with other molecules that are its targets or may rotate or spin to allow the passage of ions through a membrane. The problem of identifying the amino acid sequence, which is done by studying the genome, is simple compared to the problem of determining how this chain of amino acids, often tens of thousands of amino acids long, will fold into its final conformation. Not only is the folding a function of the orientation of atoms within the amino acids, but is also a function of the dielectrics or relative charges of some of the amino acids that can act in an attractive or repulsive fashion as do magnets. This is further complicated by the fact that the orientation is influenced by temperature and by pH level.

The determination of protein folding, even with the aid of computers is enormously difficult. Cyrus Levinthal in 1969 noted that a typical protein can have as many as 10^{300} possible folding configurations. That's the number one followed by 300 zeros. And yet these proteins, when synthesized, seem to fold into their appropriate shape in fractions of a second.

In 1994, the Critical Assessment of Protein Structure Prediction (CASP) was launched in which researchers would compete to predict protein folding, with winners announced biennially. That competition was effectively won in 2020 by Google's AlphaFold, one of the most important contributions of artificial intelligence to global health.

Some proteins are structural and are polymeric like collagen and casein and some proteins perform functions like the enzymes of rennet. Let's think about these functions in broader terms. The function performed by rennin, the cleaving of casein, can be thought of as a simple program: when attached to casein, change the configuration in a way that cleaves the molecule at a particular spot. We could express this function in programming language in not too many lines of code. When rennet is added to milk at the correct temperature, it is, of course, not a single molecule of rennet that is infused but a huge number. Each molecule of rennet is running its same little program, all of them simultaneously. In biological systems such as those involved in digestion, other enzymes are present running their own simple programs. This is a sort of computation with which we rarely have to deal: a lot of small programs running at the same time as embodied by different physical machines. In complexity science, these little independent programs are called *agents* and we would refer to this as a multi-agent system. The product that emerges is the result not only of the programming of each individual agent but of the interaction between agents.

Let's reflect back on our discussion of self-assembly in the example of the sandpile model. We saw that a consistent structure arose through the interaction of multiple grains of sand, all of which were the same. We can think of each of these grains as agents, but pretty boring ones. On the other hand, the agents of biological systems, enzymes and other proteins, are anything but boring, each of which doing its own little computation. But in the same way that the pattern of the sandpile or the attractor of the double pendulum experiment or aperiodicity of Stephen Wolfram's cellular automata were impossible to predict by examination of their elements individually, the outcome of the interactions of the multiple agents that are proteins are also often opaque until they manifest as a trait or as a pathology.

I described this book as an exploration of complexity and information in the natural world, and it took us until Chapter 4 to define information. It has taken us until this moment for us to define complexity.

In the field of complexity science, the discipline this book attempts to describe, a complex system is one composed of multiple components that are generally heterogeneous, that is to say that they are different from each other. Complex systems are not merely a collection of parts as a bag of nuts and bolts might be, but are composed of parts that interact with each other demonstrating interconnectedness and also interdependency. We can see this in biology in the interplay between proteins and other biochemicals, each of which is an agent in a complex system. The collective action of the agents is often nonlinear and unpredictable in the way that traits and pathologies can generally not be predicted simply by examining the functions of the individual proteins involved.

An oversimplified but useful example of a system that is almost complex is the game of chess. In chess, different pieces have different properties which is to say that they are capable of executing different programs. In the case of the pawn, the predominant program is to move one square forward, although a few other programs of course exist for the pawn. As the game progresses, the strength of each player's position changes until the match is over and one player has won. But there is nothing in the advantages and disadvantages each player obtains that is directly the result of the program that each of the pieces is running. Put another way, the winner of a chess match would never ascribe his victory to his third pawn from the right or from the strategy employed by his left bishop. The game is certainly dependent upon the programs run by each of the pieces, but it is the interdependency and interactions between these pieces that form the narrative that we recognize as the game.

A protein is very much like a chess piece. It is an agent in a complex system, but it is often no easier to draw a connection between a protein and a particular trait than it is to draw a connection between the king's rook and the game's victory. Truly complex systems do not consist of turns in which only one agent moves. In real complex systems, everything is happening all at once.

Complex systems can behave in unpredictable and dramatic ways, but they also demonstrate a degree of robustness. A classic example is the stock market. The stocks themselves are, of course, not agents in this game. They are more akin to the fat globules in our cheese production and are the substrate on which the agents act. The agents in this case are individual, institutional, and computer stock traders who execute their programs of buying and selling. Like biological systems, the stock market is subject to environmental influences which we generally think of as "news". Perhaps word of a semiconductor shortage, an oil crisis, or a distant hurricane change the milieu in which the market operates and the market adapts to these stresses. But it is robust in the sense that it is never the case that all of the shares of all of the companies have been purchased and held and that the market ceases to exist. The market changes in character, but it is ultimately robust to the challenges it faces.

We talk about bull markets as the price of stocks rises over a period of time or bear markets in which stock prices fall. The transition from bull to bear market is difficult to predict. One might even argue that it is not, in any meaningful way, predictable. These surprising and occasionally large-scale changes are one of the defining features of complex systems. The bull market acts differently from the bear market. It is almost as if they are different forms of the same complex system. Indeed, the analogy generally used in complex systems is that of *phase transition* as if the complex system were going from a liquid state to a gaseous state and, like phase transitions in matter, the moment of the transition can occur quite rapidly.

I have, in a few of these chapters, described a system as being "greater than the sum of its parts" and I want to define more precisely what I mean by this. Complex systems demonstrate properties that are present in none of their elements. One example that is easy to recognize, although the system is not truly complex, is the characteristic of water to be wet. I'm not going to describe wetness since any human being can recognize it, but I will ask the question how many molecules of water does it take to be wet? Certainly, one molecule of water is not wet. Wetness is a property that occurs only with a large collection and interaction of water molecules. Wetness is a property of the collective that is not present in the individual subunits.

In the case of living matter, reproduction is emergent. There is no enzyme that reproduces. Reproduction, metabolism, each of our senses is emergent. There is no substance, no agent, no protein that itself sees or hears or tastes. These are properties that exist only in the whole and not in any of its parts. Emergent properties arise without direction in the same way that our sandpile arose without direction. Our polygenetic traits are themselves emergent, which is why it is so difficult to connect the dots between the genes and the trait.

Understanding complexity is about the primacy of understanding process over product. We need the chess pieces to play chess, but it is not important of what material the pieces are made, only their role as the agents in the system.

But where does this leave us? If we cannot a priori ascertain the emergent properties of a system, does this leave us no tools to describe it? Is emergence going to be our deus ex machina and are we simply going to say that the brain is emergent and wash our hands of the rest of the matter? Of course not! Because as robust and as flexible, as surprising and unpredictable as complex systems are and the emergent structures they produce, they do not by themselves infuse additional information, only additional outcomes from information that already exists. Our inward journey has taken us very far. But perhaps it is time for us to take a walk outside.

In this chapter, we learned to see ourselves as collections of machines running simple programs and we called these machines *agents*. We discovered that interdependent agents may collectively give rise to properties that are distinct from any properties the individual agents possess and we called this *emergence*. Finally, and at long last, we defined complexity as systems composed of heterogeneous interacting and interdependent agents that collectively behave in nonlinear fashions and can produce large and unexpected changes in behavior, and we called these changes *phase transitions*. Finally, we recognized that complex systems can ingest environmental stimuli and stresses but that the system itself is not a source of information.

Chapter 10: An Impression of the World

I start this chapter with a story that, while apocryphal, is apropos of our project. Michelangelo, one of the most important figures of the Renaissance, was asked how he goes about producing marble sculptures. In some tellings of this story, the particular sculpture in question is that of David. Michelangelo is reported to have said "the sculpture is already complete within the marble block, before I start my work. It is already there, I just have to chisel away the superfluous material".

This idea captures the neoplatonic idea of the realm of the forms, in this case the form of a young human man as represented by David, that is itself perfect, and the people we encounter are imperfect shadows of the form of the human that exists only in the realm of the forms.

The idea of beginning with a block of marble and removing everything that is not sculpture has tremendous resonance for us in our ambition to build a brain.

In this chapter we will use all of these skills that we have learned in the past nine chapters to complete our task, but we will do it by changing the nature of the task itself. It seems impossible that we would be able to construct the great complexity that is the brain by employing genetic expression, self-assembly, and fractal processes with iteration of instruction, and I believe that this is true. There is simply not enough information to construct a connectome. But perhaps we're going about this the wrong way. If it is too ambitious for us to carve the sculpture, we can instead endeavor to construct the marble.

In terms of neurogenesis, what would this marble look like? It would have to be a structure that lends itself to being molded into the connectome and, like the marble, it is the structure that would need to have superfluous material. This is an achievable way forward and, amazingly, it is the one that evolution has chosen to take. To build our neurological marble, we need to make sure that we have bundles of neurons that follow paths that include those that will be part of the connectome but may also include those that are of no use to us. We can visualize this as directions to get from home to work. What we would like is a set of instructions that tells us where to turn and gives us landmarks. Generally, when we think about instructions, we start out with nothing and then build our instructions up from there. But let's take the opposite approach. Let's start out with a set of instructions, of directions, that describes turning on every possible street in every possible direction. The instruction set would be extremely large and would not be all that helpful in getting us to work. However, we can be certain that somewhere within all of those possibilities is the proper set of instructions for our commute. We can then remove those instructions that do not get us to work, to leave only those instructions that describe the actual appropriate path.

The set of instructions that includes every possible turn in every possible direction on every possible street is the marble and we need to be Michelangelo and simply remove those instructions that are not our commute.

Is this really the way that brains are built? Is there really an overabundance of connections that are then carved away to reveal the sculpture that is the finished connectome? Well, if that were true, we would see the bizarre and counterintuitive case that the baby has more neurons than the adult. Surely, this cannot be true.

And yet it is. At birth, we have about 100 billion neurons. As we grow to adulthood, an apparently superfluous 14 billion are carved away. This reduction in neurons is not equal in all parts of the brain but is especially prevalent in the neocortex, that portion of the brain involved in what we think of as higher functions like reasoning and language. If we look at synapses, the numbers are even more dramatic. By about age two or three, we have our peak number of synapses at about one quadrillion. As we mature, between a third and a half of these synapses are removed.

In this context, the removing is the result of *apoptosis*, what is often called programmed cell death. Although difficult to imagine, a sort of apoptosis is evident to anyone who has ever had the fortune to play with a preverbal child. John McWhorter, an Associate Professor of Linguistics at Columbia University, asks us to examine beatboxing, the musical performance in which the beatboxer emulates the sounds of musical instruments and environmental sounds in an entirely acapella fashion. McWhorter makes the obvious point that the beatboxer can only make sounds which can be made, and points out that many of these sounds are sounds that we ourselves made when we were infants. A very young preverbal child will make all sorts of noises with the mouth and the throat, and those noises that are recognized by the parents as components of language are reinforced. The reason that the Khoisan can include clicks in their spoken language is because these sounds were reinforced during their early formative years. The reason that you and I cannot make these sounds is because, like many other sounds, they were pruned by a sort of apoptosis of diction. Successful beatboxers are perhaps rediscovering sounds that they themselves made as infants.

The idea of overproduction of neurons and synapses with subsequent cell death sounds enormously wasteful but it is wasteful only in the sense of material and not in the sense of information. In the same way that our fractal snowflake can have infinite intricacy but very little information, the infant brain can contain more neurons but less information than the adult brain.

We have our marble in the form of the overabundance of neurons in the infant's neocortex and synapses in the toddler's neocortex and now we have our chisel in the form of apoptosis. What we are missing is our sculptor.

I have divided this book into three parts as will surely now be apparent. Chapters 1 through 5 dealt largely with genetics and compression and an examination of the information content of the genome. Chapters 6 through 9 dealt with complexity and self-organization. Chapters 10 through 14 are about the synthesis of these two sets of ideas.

The exploration of complex adaptive systems was interesting, but in the end, we decided that these complex systems do not add any information, but rather add an additional way of dealing with existing information. I want to make the argument that in the adaptive nature of these systems, we will find our sculptor.

Let's return to our example of the stock market. When the price of stocks in the technology sector is rising and a new bit of news arrives that causes the price of stocks in this sector to fall, this sector undergoes a phase transition in the sense that the sector acts differently before and after the news story has broken.

Viewed from the perspective of information flow, the market has ingested information and has become changed by it. It is almost as if this information has left a metaphorical impression on the system that has changed its shape. If another news story arrives that also transforms the behavior of the market, the influence of the first story is not completely gone. The system has ingested a second story and the system will continue to change as further news arrives. The system does not forget the old news even as it alters itself to accommodate the new. The complex system ingests and adapts and this is the result of the interaction and interdependency of its parts.

The brain does the same. If we remember back to Chapter 7 in which we talked about concentration gradients and chemical signals that direct the growth of neurons, it will come as no surprise that these signals are also employed during the process of apoptosis, the development of synaptic weights, and neuroplasticity. This production is cued by ingestion of information from the environment. In the case of refractive amblyopia, lazy eye that arises from the uncorrected requirement for glasses in just one eye, too little visual information is ingested as a result of a poor quality image being projected onto the patient's retina in the eye that will become lazy. This lack of information affects apoptosis in a way that prevents the reinforcement of connections in the visual system and cannot be overcome by later neuroplasticity. It is an example of exogenous information directly affecting anatomy and brain function. It is evidence that some of the information for construction of the connectome arises from outside of the organism, that is to say, from the world at large.

Let us take the simplest example of which I can think. When I was a young child, the teacher would take us to a church with ornate carvings and would give us paper and crayon. Rather than asking us to draw the figures represented in the carvings, a task at which I was especially poor, we were told to lay the paper over the carving and to gently rub the crayon over it. This rubbing would produce a facsimile of the bas relief and we would then see a picture appear on our papers. Let's examine the parts of this highly imperfect analogy. We can think of the paper as the brain, the crayon as apoptosis, and the process of rubbing the crayon over the paper, laying down thicker lines where the bas relief rose up and lighter lines where the bas relief dipped, as the adaptation that is characteristic of complex systems. We wind up with the art transferred to the paper. We understand the role of the paper, the role of the crayon, and the role of the process of rubbing. But from where did the information for this picture come? It did not come from the paper nor from the crayon nor from the process of rubbing. The information came from the external world.

In this case, the picture on our page represents a literal impression of the world, but in the case of apoptosis and its role in development of our brains, it is a metaphorical impression of the world. But in both cases, the rich source of information is external to the object at hand.

In chapters 1 through 5, we built our marble. In chapters 6 through 9, we built the system of rubbing in that we observed complex systems adapt to their environments. What is adaptation if not the ingestion of information and the reformation of the system to accommodate this information?

This is a profound insight in many ways. In Chapter 8, we talked about von Neumann machines and the idea that such a self-replicating machine would require not only the instructions for its construction but also a system to execute that blueprint. We are not built that way, at least in the development of our neocortex. Our system offloads the burden of much of the information of our construction to the environment itself. The brain cannot be built in isolation of its external environment.

In another example, in the field of language acquisition, I will relate the sad story of a language deprivation experiment that Herodotus recounts in his histories of the Egyptian pharaoh Psamtik. From a modern English translation of Herodotus's Histories, book 2:

Now before Psammetichus became king of Egypt, the Egyptians believed that they were the oldest people on earth. But ever since Psammetichus became king and wished to find out which people were the oldest, they have believed that the Phrygians were older than they, and they than everybody else. Psammetichus, when he was in no way able to learn by inquiry which people had first come into being, devised a plan by which he took two newborn children of the common people and gave them to a shepherd to bring up among his flocks. He gave instructions that no one was to speak a word in their hearing; they were to stay by themselves in a lonely hut, and in due time the shepherd was to bring goats and give the children their milk and do everything else necessary. Psammetichus did this, and gave these instructions, because he wanted to hear what speech would first come from the children, when they were past the age of indistinct babbling. And he had his wish; for one day, when the shepherd had done as he was told for two years, both children ran to him stretching out their hands and calling "Bekos!" as he opened the door and entered. When he first heard this, he kept quiet about it; but when, coming often and paying careful attention, he kept hearing this same word, he told his master at last and brought the children into the king's presence as required. Psammetichus then heard them himself, and asked to what language the word "Bekos" belonged; he found it to be a Phrygian word, signifying bread. Reasoning from this, the Egyptians acknowledged that the Phrygians were older than they.

What actually would happen to such children? Documentation of cases of children raised without being spoken to or hearing any language reveal them to remain cognitively deficient. Without the ingestion of exogenous information, the construction of the functioning brain cannot be complete.

And certainly, the information flow is bidirectional. It is not merely that the child absorbs stimuli from the world but, in interacting with the world, the child can cause the world to generate further stimuli. When the child touches or manipulates an object, when the child vocalizes and then hears the vocalization, the child is providing an input to the environment and then ingesting the information that arises from the environment as a result. We say that the child is learning about the world, but it is far more profound than that. The child is, as we are, part of the world and a large part of the instructions for building us is not found within us at all.

In this chapter, we finally came to terms with the fact that the genome is simply insufficient to build an adult brain, but we built one anyway. We did this by recognizing that complex adaptive systems are information ingestion engines and that it is a simpler task to create these ingestion engines and to build the substrate on which these complex systems will act than to design the final product. As with our very first chapter, we see arise again the process/product dichotomy and we see the singular importance of exogenous information.

It is at this point that I wish to pivot the entire project of this book on its head. The environment, the world, is indeed a rich source of information. But although this has served us greatly in our quest to construct a brain, it also presents a crisis. We have gone from having too little information to having far too much. How can we with our enormous but finite brains operate in a world in which information is essentially limitless and is far too much for us to ever understand? This crisis of reality will be the subject of our penultimate chapter and I can't wait to discuss it with you.

Chapter 11: The Crisis of Reality

My son's dog, Dishsoap -- yes that is his name -- is afraid of boxes. Apparently, for dogs in New York City this is not an uncommon fear. As an aside, New York City dogs are a special sort and, being a golden cavadoodle, every other dog in New York looks like Dishsoap's clone.

Dishsoap is not in the forefront of thinkers of the canine world, but I think that makes this point even more clear. Why is Dishsoap afraid of boxes? Has he ever witnessed a box eat a dog? It seems to me that Dishsoap has assigned both agency and intentionality to the box. That is to say, in his worldview, in the model of the world he has constructed in his head, the box is capable of doing things and the box may want to do something that would be unpleasant to Dishsoap.

I think I can be fairly confident in saying that Dishsoap is afraid that the box will perform an action that would result in unpleasantness. If in fact this is what's going on in his head, and of course I am open to reasonable alternatives, then Dishsoap must have at least some rudimentary sense of causality in the form of box → action → harm.

We are susceptible to the same sort of cognitive biases, assigning agency where none exists, e.g. "this computer hates me", and assigning intentionality where there can be no intention. We even have a special word in our language for this and that is anthropomorphism.

I want to make clear that I am not discussing the idea of theory of mind in dogs or other animals. However, I am saying that the understanding of intention has generally been associated with higher cognitive function. I want to suggest the opposite. I hope to make the case that these cognitive biases, and others that I will discuss, are evidence of cognitive insufficiency in the face of the richness and complexity of the world.

It is impossible to conceptualize the complexity of the world in its completeness and depth and we explored this idea when we discussed Laplace's demon. Instead, we say we focus on "what matters". In the language of cognitive neuroscience and evolutionary psychology we are decreasing our "cognitive load" to one that is workable. Thinking back to our very first chapter in which we discussed compression in the genome and said that the compression had to be lossless. We are dealing with a sort of compression here as well when we're trying to reduce the real time calculations that would represent the world into a smaller set that is more manageable but still sufficiently representative. In this case, we are going to employ lossy compression in that there will be some information that we simply discard. Some of this is obvious to us and as illustrated in idioms like "not getting lost in the weeds." But some of the compression is so ingrained that we no longer see it and it is to this that I want to call our attention today.

Now, at this point, we understand that changes and transitions like growth and the stock market and the weather represent evolutions of complex systems involving the interaction of a multitude of agents. We cannot point to a single agent or the process it executes as being the cause of the transition of the system, but that rather this transition occurs because of the interaction and interdependency of the individual agents.

However, sometimes a single agent is of such a magnitude that its influence subsumes those of all other agents. When a hunter shoots a deer, the influence of the agent that is the bullet on the complex system that is the deer is so overwhelming in magnitude that we can essentially ignore the contributions of all of the other agents that make up the deer: the enzymes and biochemical substrates of Chapter 9. We can approximate the evolution of the system by simply saying the hunter killed the deer with a bullet and this will be so close to reality as not to matter. This construction of A causes B is what we call simple linear causality. And the transition of a system as a result is what you and I call "an event." We assign to this event a *beginning* and an *end* and this description that takes place over a discrete amount of time is what we call a *narrative*.

Linear causality can involve many steps as in the case of the proverb for want of a nail the shoe was lost, for want of a shoe the horse was lost, for want of a horse the rider was lost, for want of a rider the message was lost, et cetera. This chain of causality has many intermediary steps, but I want you to observe that it remains linear. This is quite distinct from the complex systems we have been describing in this book.

Neurodevelopment, although it evolves over time, is certainly not linear or at the very least not serial. It is a highly parallel process involving many parts influencing each other over time, and, as such, is representative of what actually takes place in the world. In complex adaptive systems, unless the system truly collapses, there is no beginning or end. There is no possibility of narrative in the conventional sense. There is no simple causality. The construction of a narrative and the assignment of linear causality are approximations that we use to reduce our cognitive load. In some systems, as in the case of the deer, a single agent can be so overwhelming that the transition can be approximated with a narrative. But the fact that we recognize this as an event is, in a sense, the exception that proves the rule. We do not perceive the world as a series of constant and overlapping events because, if we did, the idea of an "event" would have no meaning. Events are, by their nature, exceptional. It is not that nothing is happening between events. It is that what is happening does not lend itself to narrative or causality and can generally be ignored without imperiling us.

The ideas of narrative and linear causality are expressed in other ways in our lives as well. Have you ever been so frustrated with your computer that you've wanted to yell at it? If you were to be asked about this impetus, you would certainly not suggest that you truly believe that the computer is out to get you. But why is this such a universal response? By assigning agency to the computer, we can construct linear causality that fits into our world model. The computer hates me and has made my life difficult. Again, I am not suggesting that any of us believes that computers have agency, but these are such universal feelings that they represent a sort of cognitive bias. Would such a bias serve any selective advantage?

Kurt Gray and Daniel Wegner in their paper of 2010 wrote, "The embarrassment you feel hopping out of the water after mistaking a wave for a shark is nothing compared to the pain of having your leg eaten after mistaking a shark for a wave. The high cost of failing to detect agents and the low cost of wrongly detecting them has led researchers to suggest that people possess a Hyperactive Agent Detection Device."

Some problems, like avoiding a predator, require rapid processing and this sort of bias, of simplification, can be helpful. Indeed, suggestions of causal reasoning have been observed in corvids, a group of birds, as well as in non-human primates. The idea of causality requires some sort of model of the world and presumably those of corvids and non-human primates are not as sophisticated as ours. But it seems that some of the same shortcuts have been taken.

All of this goes to say that it is enormously difficult for us to internalize the idea of complexity. I certainly cannot do it. The approximation that is our understanding of causality is so ingrained that we no longer see it. Indeed, our language, the very words I write, consists of subjects and verb phrases and themselves represent a description of linear causality. As a human being, I can imagine no other way. I do not know what a language of mutually influencing parallels would look like.

These biases can, however, lead to problems of interpretation. When we say that smoking causes cancer, this simplification of a complex system is sufficiently accurate that we can incorporate it easily in our worldview. However, when we say that the latest weight loss diet fad will make us skinny, our biases towards linear causation do us a disservice. Even things like changes to cancer risk by exposure to low level radiation are difficult for us to interpret, and we talk of changing probabilities of cancer rather than of certainties. But what do these probabilities mean? Do they mean that there are stochastic elements involved? Certainly, this is the case. But I would put forth that even if there were no stochastic elements, even if we ourselves were entirely deterministic, there would still be no certainties, only probabilities. This is akin to the deterministic chaotic systems that we saw in which the outcome was unpredictable even though all of the rules of all of the elements were known. Each of us is a slightly different complex system and these subtleties, these initial conditions, make forecasting difficult and sometimes even impossible.

I am a scientist and I say, without equivocation, that science is the means by which we will gain understanding of the universe. But I do want to point out the influence of narrative and causal biases in the scientific method. In science, we make observations and form a hypothesis that we then test. The difficulty is that many of these hypotheses have their own narrative biases. It is rarely the case in biology that A causes B. Better, it is to understand that A influences the system in a way that makes B more likely. The difference is not semantic but genuinely conceptual. And it is difficult.

How might we move forward, as scientists, with our understanding of complexity? One model for us might be one influenced by the approach of physicists. One area in which multiple interacting agents influence the state of a larger system is in the field of quantum field theory. Now, before I go any further, I am not suggesting that any of the systems that I have described are quantum and I am not going to explore the role of quantum physics in neurology. I am merely presenting the conceptual framework of quantum field theory as an example of a highly parallel interactive system.

When we speak of fields in physics, we are describing particular properties that are distributed over space. One example with which we are all familiar is of magnetic fields. In quantum field theory, fields are influenced by the presence of particles such as photons and electrons. Before my physicist friends tune out, I know that my last sentence is untrue and I will revisit it in a few paragraphs.

Each of these particles has properties and the particles interact with each other according to their properties. This is beginning to sound, very generally, like our description of complex systems. In the case of electromagnetism, two particles interact by exchanging a particle that represents the electromagnetic force, in this case, a photon. The classic analogy is of two skaters on an ice rink throwing a ball back and forth. The ball itself conducts some amount of momentum from one skater to the other. In a similar way, the photon is a vehicle for the electromagnetic force between two particles.

In truth, quantum field theory does not view these agents as discrete, but rather views all of this as features within a single quantum field. The electrons themselves are probabilistic perturbations of the field, but the view is much more holistic. The field itself evolves over time according to Schrodinger's equation. The complex biological systems we've discussed in this book are analogous to the field and, in the same way that the quantum field is adaptive to perturbations from electrons and other particles, our complex biological systems are adaptive to environmental influences. So far so good.

Carlo Rovelli, a physicist and author with special interest in loop quantum gravity, describes the objects in the field, and to some extent all objects, as collections of properties that describe the nature of the interactions they have with other particles. To Rovelli, the electron is not so much a discrete physical object but is rather defined by *what it does* in relation to other particles. The idea that something *is what it does* is another example of defining a system as process rather than product, one of the chief themes of this book. In the case of rennet in cheese making, of a chess piece in a game of chess or, indeed, of the little machines we call proteins, what is important to the system as a whole is what these agents do and not of what they are made. If an enzyme were swapped out of this system for something else, let's say a bit of nanotechnology that had identical properties, the system as a whole would not change at all. In this way, the methods of quantum field theory can serve as a guide for us in interpreting complex systems in the macro world.

There are problems with this approach. All electrons are the same. In biological systems, individual agents are often different from each other, and these differences introduce important stochastic elements that are absent in quantum field theory. This is, of course, not to say that quantum field theory does not have stochastic elements. But if you've seen one electron, you've seen them all. There is a small literature in the application of field theory to complex systems and this may be a fruitful approach going forward.

In this chapter, we have discussed the impossibility of incorporating the richness and complexity of the world into a model we can fit within our limited cognitive resources. We discussed the ways in which evolution seems to have reduced cognitive load by producing simple models that are approximations of reality. We saw how these approximations lead to biases of agency and intentionality and a bias towards understanding the world in terms of narrative and linear causality. We observed that other fields within science have had analogous challenges and have met them with a holistic approach such as in the case of quantum field theory. We have seen that our quite understandable difficulty in conceptualizing the tremendous parallelism and simultaneity of reality has shaped the model of the world that we hold in our own brain. But what does this mean for brains that we build? What are the implications for artificial intelligence? This will be the subject of the twelfth chapter of this book and I look forward to discussing this incredible subject with you.

Chapter 12: Artificial Intelligence

I start today's chapter with a story, a gedankenexperiment, a thought experiment that has been elevated almost to the level of a parable in AI circles. It's the story of a very special room published in 1980 by John Searle of the Department of Philosophy at UC Berkeley. The story has many variants even within the original paper, but I will present the simplest here. Inside of this room are three books and a native English speaker who understands and reads no Chinese. In the door to this room there is a slot for messages coming in and out. Inside of the room is the table with three books. To the person inside the room, these three books look very similar. In each book there are squiggles and lines with associated reference numbers. The person in the room is also given a set of instructions written in English with rules to match a reference number in book #1 with a reference number in book #2 and similarly a set of rules to match a reference number in book #2 to a reference number in book #3.

The person in the room has a job. When he receives a piece of paper through the slot in the door, he carries it to book #1. The paper is filled with slashes and squiggles and other similar shapes, and he goes through book #1 to find the slashes and squiggles that match those on the paper. He then follows the first set of rules (remember these are written in English) to match what he finds in book #1 with what he finds in book #2 and then another set of rules to match these with book #3. Finally, he copies down the squiggles from book #3 in order on his sheet of paper and passes the sheet of paper out the slot in the door. He is proficient at this job and spends his days receiving pieces of paper and pushing pieces of paper he has written through the slot in the door.

On the other side of the door is a native Chinese speaker who writes questions about a particular story onto a sheet of paper and passes them through the slot. He then receives answers to his questions about this story written onto sheets of paper and passed back through the slot to him.

The designers of this system have a special jargon for the three books on the table, the sheets of paper going in and out of the slot, and the set of English instructions the person in the room is given. The three books are referred to as script, story, and answers; the pieces of paper are, of course, called input and output; and the rules written in English are known as the program.

Let's reflect upon this for a second. The Chinese user obviously understands Chinese. But in the black box that is the Chinese room, who understands Chinese? It is not the man in the room whom we've already described as not knowing a word of Chinese written or spoken. Who then? Does the room understand Chinese? And if the man in the room does not understand Chinese, how is he giving answers about the story? Who understands the story? Is it again, the room? Do the books understand the story? Do the rules speak Chinese?

This was a fascinating thought exercise and was a lot of fun to debate until November 30, 2022 when ChatGPT was launched. Now each of us has his own Chinese room.

Artificial intelligence encompasses a large number of approaches and I want to confine our discussion to algorithms that require training. It's difficult to imagine any other sort of artificial intelligence now, but you'll remember we discussed symbolic logic and Marvin Minsky and the AI winter that resulted in part from Minsky's criticism of neural networks.

When we think of artificial intelligence, we think of algorithms that require training. Convolutional neural networks can be trained on a series of images in order to distinguish shoes and watches and hammers or to distinguish different sorts of pathology that are identifiable on radiologic, micrographic, or optical tomographic images.

The pixels of the image are arranged, often linearized, and given as input to the neural network. The neural network does calculations on these pixels and then categorizes the image. Its accuracy in categorization is assessed and variables within the algorithm are altered based upon the accuracy of the output of the neural network. Other sorts of machine learning work in analogous ways.

The training data consists of images or other sorts of data that have already been labeled by humans with the correct characterization. For example, the training set might contain pictures of lions and of dogs with the pictures of lions having been labeled by humans as "lion" and the pictures of dogs being labeled by humans as "dog". We refer to these labeled data as real world evidence or "ground truth" and we assess the ability of the AI algorithm to assign the correct label to the images it is given, then alter the variables within the algorithm to guide it to the correct answer. This is called training.

The algorithm is then exposed to another set of labeled data, of ground truth, that it has not seen before. In the parlance of AI, we say that we are showing it data to which it is naive. We then grade the algorithm on its ability to correctly label this new data set and, based upon how it does, we can say whether the algorithm performs well or not. We call this stage validation.

This, by now, should be very reminiscent of apoptosis and neuroplasticity in which an initial architecture is built and then is modified based upon exposure to outside data. In the abstract, we can say that the algorithm is developing a model of the sort of data to which it is exposed. This model is limited to the subject matter on which the algorithm is trained. Algorithms trained to detect prostate cancer develop internal models only of prostate cancer images. They have nothing to say about the stock market or neurodevelopment. For the sake of this discussion, I'm going to refer to the algorithm's model of the data on which it is trained as a "world model", understanding well that this world is generally very limited.

This has some parallels to our own neurodevelopment but also some important differences. The only interaction between the algorithm and the data is one-way. We acquire some of our information in a one-way manner, but our foundational knowledge is more interactive. We learn how to manipulate objects, indeed we learn of objects' properties, by handling them as infants and toddlers. We learn through our senses in a largely bidirectional manner. Even our senses of smell and of vision which seem to be entirely unidirectional are, of course, highly modified by our movement through space. As we walk around, the image obtained by our eyes, even in a static environment, changes. The same is true of smell and hearing. It is through this two-way interaction with our environment that synaptic weights get changed.

There are several profound implications that arise from the simple fact. One of them is that our neurodevelopment is inextricably linked to our senses. In his seminal paper of 1980, Thomas Nagel asks the question "What Is It Like to Be a Bat?" Nagel makes the point that it is impossible for us to understand a world that is perceived through echolocation. That is to say, the sensory apparatus is strongly linked to the type of intelligence that is built with its assistance. Much ink has been spilled on discussions of sensory experiences or qualia and these are of great importance in consciousness but, I'm afraid, this book is not the right forum for that sort of discussion. In the much more atomistic analysis in which we have been engaged, we can say that the world model of the AI is entirely dependent upon the sort of data on which it is trained and if that data is echolocation rather than photographic, the AI will build the world model based upon echolocation.

If AI is built upon data, and data is acquired, ultimately, through sensors, would it make more sense to incorporate sensors into the AI itself? Indeed, this is a fertile area of research in the field of robotics and is referred to as "embodied intelligence". In practice, a traditionally trained AI model is built, and then additional modifications are made employing what is called reinforcement learning through sensory interaction between the embedded AI and its environment. This parallels the brain development we have discussed in which some neurologic processes are prebuilt in those areas that are sometimes called our "reptile brain" and other areas such as the neocortex are greatly modified by interaction with our environment (see Chapter 10). This can result in a much broader world model than is usually obtained through traditional convolutional neural networks. But, I want to stress again, that this world model is constructed out of sensory data and if that sensory data were echolocation or optical polarization, or detection of magnetic fields, a facility present in some birds, then the world model will be entirely alien to those constructed in human neurodevelopment. Ultimately, it is impossible to separate an intelligence from the world in which it is built. One of the lessons of Chapter 10 is that neurodevelopment is inseparable from the outside world. One of the corollaries of this chapter should be that the brain is inseparable from the body and its senses.

But what of large language models like ChatGPT and its ilk? They seem to know everything! To understand what's going on here, we need to return to our Chinese room. Remember, the Chinese room consisted of books, a non-Chinese speaking person, and a set of rules that told the person how to gather information from the books. We asked the question, who in the system knows Chinese? Is it the books, or the rules, or, unwittingly, the person? The answer to this question is that none of these entities knows Chinese. When you use a calculator, do you believe that the calculator knows mathematics? Of course not. The calculator has a set of rules that it executes to produce an answer and, to the extent that we can say that the calculator knows anything, it knows these rules.

Similarly, in the Chinese room, the translation is not done by the books, which are a static database, or the person who is the CPU in this system, but is rather done by the set of rules or the "algorithm". I don't think that there is any cognitive leap here in saying that the algorithm is performing the translation. But that is quite different from saying that the algorithm *knows* Chinese. The real questions are about how this algorithm was produced. One way that we know to produce such an algorithm is through the training of an AI. This training involves exposing the AI to labeled data and then modifying the AI until it gets things right. The people who know Chinese are the people who labeled the data on which the AI is trained. So the answer to who in the system of the Chinese room actually knows Chinese, is the people or person who produced the data on which the algorithm was trained. It's really not that much of a mystery.

In our earlier example of a convolutional neural network able to distinguish dogs from lions, we can ask: who actually knows about dogs and lions? The answer is the humans who labeled the initial training set. The algorithm has developed a world view consisting of three things: dog, lion, and other. We cannot say that the algorithm understands what a dog is any more than we can say that the algorithm of the Chinese room understands Chinese. But it is able to categorize objects within this greatly simplified world model.

What about the large language model (LLM)? ChatGPT seems to know <u>about</u> lions and <u>about</u> dogs. Does this mean that the LLM has a world model that encompasses all of the objects of the actual world? To explore this question, we need to ask, as we did in the case of the Chinese room and in the case of the lion and dog algorithm, what is the nature of the training data? The training data for large language models is vast and consists of much of what exists on the Internet. Most of this, initially all of it, consisted of text written by humans. These humans, as do all living organisms, exhibit embodied intelligence. We learn about our world by being in it. We can then describe the world, if we are adept at composing written language, in the form of text. The text is therefore at least one step removed from the initial sensory data. And it is this text on which the LLM is trained. The LLM's world model is not a model of the world as we perceive it but a model of the world as we *describe* it. The LLM cannot predict anything of the world itself. It can only predict description.

Let's try to make this more concrete by reexamination of the Chinese room. If we were to build 1,000 or 100,000 Chinese rooms and then collect the algorithms from each of these rooms and train an AI on those algorithms, that AI would be akin to an LLM. That AI would not only be good at translating Chinese but in creating a world model of the sorts of rules required to translate Chinese. The LLM could discuss the particular ways the Chinese language describes things and, if the LLM were also trained on 100,000 English rooms and 100,000 Swahili rooms, the LLM could discuss in detail the ways in which the different languages divide things like color and taste and motion and other ways in which languages can differ. But the LLM does not understand these languages and has not built a model of the world in the way that we do. It has built a model of the rules of the rooms.

What is the distinction here? What is the difference between having a world model that includes a Chinese description of broken glass and having a world model that incorporates broken glasses? Certainly, to the user of the LLM, it can be very hard to distinguish the two. But there are difficulties in generalizing the world model the LLM has created outside of its training set and there are issues as well in understanding causality, and this is because the descriptions of the world are based on text that has already been filtered through our own sense of what is relevant and what is not.

In the room in which I sit, my computer beeps, there is heavy machinery outside, the room is warm, there are footsteps on the stairs, my walls are yellow, my coffee cup is empty, the surface of the desk is chipped, my speakers are dusty, it is Tuesday just after noon, and the mirror is tilted. Almost none of this information is relevant to each other and most writing, even very descriptive writing, is not as laden with irrelevance as the information that flows through our senses. We have already filtered out irrelevant detail when we write our text.

The result is that LLMs are very bad at determining what is relevant. If we give the LLM the description of the room I have mentioned, it will connect all of the parts into a single narrative and will discard nothing. This highlights the point that the LLM does not have a world model of the actual world but simply has a world model that consists of our descriptions. The LLM is very good at predicting descriptions but has no first-hand knowledge of the world through its own sensors and no sense of association, relevance, or causality.

In order to overcome this, there are research efforts to allow LLMs to interact with the world. For example, some AIs are trained to ask questions and then to employ reinforcement learning to adjust weights in their models in order to incorporate the information generated by this interaction. LLMs have also begun to incorporate the ingestion of pictures. These pictures are a sort of primary sensory data. After all, I believe I know what dinosaurs look like and yet I've never seen one myself. My information comes from pictures too! However, because of our interactions with the world, we are able to generalize information we obtain from pictures a great deal better than can computers.

I have created very compelling slides using AI-generated animation of still pictures. But some of the motion that AI produces from these pictures is physically impossible and reveals a lack of understanding on the AI's part. By manipulating objects, by walking through space and seeing how objects in the world change as I move around them, I can generalize and imagine what a view of a dinosaur would be even if I have never seen that particular perspective in a drawing.

Perhaps embodied intelligence in the form of robots that can move around in real space and manipulate objects will allow such generalizations in the future, again, emphasizing the ultimate inseparability of an intelligence from the physical form in which it is embedded and through which it senses the world.

I have two other points to make about LLMs. The first is that, trained upon the writings of humans, they incorporate all of the biases we discussed in Chapter 11. I know that I have just criticized LLMs for not understanding causality but, you will recall linear causality is not the way much of the world works. LLMs mimic text that describes a world of objects rather than of processes and largely ignore the rich interdependence of the agents of which the world is composed.

My final point about LLMs and almost all artificial intelligence is that it itself is built around a linear process. It is a program and, as a program, does its calculations in a linear order. In AI, this order may have iterative loops, there may be feed-forward and feed-backward, and it may have elements of parallel calculation. But ultimately, it is a single process. As wonderful and complicated as these AIs are, they are not complex. They are massively complicated agents, but they are, in the end, unitary entities that do not interact with each other. The best that such AIs will be able to achieve is a mimicry of human understanding of the world.

An alternative approach is agent-based modeling in which different algorithms interact with each other. It is here that we can expect to see new emergent properties that are not present in the algorithms of the individual agents. The degree of computation required to model the interaction of multiple interdependent algorithms is mind boggling. My AI colleagues can attest to the fact that the speed at which algorithms can be trained is largely dependent upon computational resources. Each simple bacterial cell can produce thousands of different enzymes and a multitude of copies of each. Even if each of these enzymes runs a simple algorithm, the real time computation is enormous. And this is for a single bacterial cell! Eukaryotic cells are much more complicated and the infant's brain contains 100 billion neurons to say nothing of glial cells and other tissue with which the neurons interact and interdepend. It is no wonder that the Human Brain Project found itself in over its head.

Efforts do exist to incorporate the ideas of complexity science into AI architecture and as computation technology continues to blossom, we will one day have the resources to create systems that will demonstrate power and flexibility and, perhaps most importantly, emergence. We are living in a very special time.

Chapter 13: Microintelligence

Benjamin Franklin is my favorite of the American founding fathers. Not only did he make great contributions to science, but he was also hilarious. Notable quotations include "three can keep a secret if two of them are dead" and "he that falls in love with himself will have no rivals". Not only did Franklin have a passion for wine and, well, passion, he created an accessible library in Philadelphia, albeit with a subscription fee. I think it would be fair to say that anything that interested Franklin would interest us. And one thing that very much interested Franklin was a chess playing machine called the Mechanical Turk.

Built in 1770, the Mechanical Turk was expert at playing chess and captured the attention of not only Benjamin Franklin but of Napoleon Bonaparte as well. The device was so successful it toured Europe for 84 years until being destroyed by fire in the museum in which it was housed.

The Mechanical Turk was a device that could think and plan all before the invention of any logic circuits or, indeed, even formal Boolean logic. It was an entirely mechanical computer. Or at least it would have been, if it had not been a hoax. Sometime after its destruction, the Mechanical Turk was revealed to contain a chess playing person who would operate levers to give the illusion of a thinking machine.

In what can only be described as the 21st century equivalent of envelope stuffing, Amazon launched a service to provide ground truth image labeling for AI development and named it Mechanical Turk. Now you can have a side hustle labeling photographs and to serve as the human in the machine.

Does the revelation of this hoax mean that devices without circuits or neurons cannot think? In order to avoid the morass of defining thought, for the purpose of this conversation, let us define thought as universal computation such as is done by our computers and such as was defined by Alan Turing. Indeed, such entirely mechanical computers were produced, or at least designed, in the early 1800s. Charles Babbage received a grant of £1700 by the British government in 1823 to design a mechanical computer. His original calculator, although not a general-purpose computer, was hand cranked. His work on an analytical engine, which would have been capable of general computation, would have, if built, consisted of 25,000 parts and weighed 4 tons.

Of course, living organisms do not perform computation with room sized collections of mechanical parts. Computation is performed electrochemically by means of neurons and associated structures. But in the same way that electronic circuits are not necessary to build computers (although, it would seem that they are necessary to build *practical* computers), are neurons necessary to build organisms capable of performing computation?

One of the simplest multicellular organisms is the sponge. Composed of only a few cell types, sponges have no nervous system. Sponges spend their sessile lives cemented to the sea floor and aside from filtering ambient water, generally don't do very much. In such a setting, one might think that no computation is necessary. However, this turns out not to be the case. Because even though the seabed may not change much, ocean currents do. In order to optimize the filtering of sea water for organic material, the means by which sponges "eat", they need to adjust to changes in water flow.

These adjustments to local currents are both active and passive on the part of the sponge. The macrostructure of the sponge is one that alters the flow of water in a manner that facilitates food capture. On a smaller scale, the choanocytes of sponges, those cells that incorporate flagella to actively move water, can change their behavior to accommodate local environmental changes. On the level of the individual cell, this may not seem particularly remarkable, but these adjustments and activity appear coordinated, and this is done without the harness of a nervous system.

The idea that structure itself can perform computation is, at least to me, viscerally unsatisfying. By this measure, the architecture of any building with sloped roofs or elevated chimneys can be said to exhibit structural computation. Even though this is true, it does not feel like genuine computation. Indeed, the example of sponges begs the question "what is computation?"

Computation turns out to be a difficult concept on which to form a consensus. The subject is wrestled with brilliantly in the Stanford Encyclopedia of Philosophy. As with every other word in the English language, the meaning of computation seems to evolve over time. In 1954, Andrey Markov, an early contributor to what we think of now as artificial intelligence, insisted that a computational process must include three elements. It must be determined, applicable, and effective. Determined for Markov meant that the instructions that are the algorithm of the computational process do not allow for arbitrary choice. In the language we have been using, the instructions are determinative. Of course, by this definition, neural networks in which the initial weights are chosen randomly might not meet Markov's definition of a computational process. This idea strikes me as ridiculous. By applicable, Markov meant that the computational process had to accommodate inputs and by effective Markov meant that the computational process must result in some change as a result of ingesting the inputs.

Nothing in this definition requires the presence of digital computers nor does it exclude biological systems from engaging in computation. Physical computation, that is to say computation without circuitry, is a good deal harder to define. Let's return to the example of the sponge. If the choanocytes receive information in the form of currents or, let's be more granular and say the vector of local water pressure over time, and then modify their behavior in a consistent way, we can feel comfortable saying that these sponges meet the requirements of Markov to be determined and effective. The system is applicable in the sense that the inputs relate to the change the system makes, but the question of information storage is not apparent in the case of the sponge. If information in the form of changes in water flow is ingested by the system, where is that information stored? Without neurons or circuits, it becomes difficult to understand what memory actually means. If the ingestion of information results in a physical change, this change can be said to itself represent memory. But that is a very broad definition. If the trunk of a tree has been marred with an axe, is the scar on the trunk a sort of memory of the trauma? I am not talking about a metaphor. Is that scar an actual memory in the sense that it can be referenced later on? In the case of the tree trunk, there is no effective change that would meet Markov's criteria for computation. But perhaps we should not be so hasty. Does the tree not heal injuries? Is this healing process not the effective element required by Markov? If the response to the injury is generally predictable, this meets Markov's definition of a determined system. Does this mean that the tree is itself performing computation?

It is very easy to so broadly define computation as to see it everywhere. There is a school of thought called pancomputationalism but, at least for me, it defines computation so broadly that the idea of computation begins to mean nothing.

Therefore, let us stick to the idea of computation as consisting of two parts: the ingestion of information and a response to this information that is at least generally predictable. In biological physical computation, we can expect the response to be somehow related to the information stimulus, but this requirement is not strictly necessary. To be a bit more concrete, we expect that the sponge's response to changes in local currents will affect the flow of water within the sponge in a way that optimizes its project of being fed. We do not expect this sponge's ingestion of information of water flow to, for example, change the sponge's color. But were it to do so, this would still constitute computation.

Plants are indisputably beautiful. Some of this, not to be too teleological, is because many plants have designed themselves to be beautiful since we, as animals, are often the means of distribution of their pollen or their seeds. But as visually compelling as many plants are, most of the action is going on underground and out of our sight. The root systems of plants are never static and adjust to local environmental conditions as well as to the changing needs of that portion of the plant that is above ground. These root systems can demonstrate remarkable degrees of computation.

We have seen that some degree of computation is demonstrable in simple animals as well as in plants, but what of organisms that are neither? Indeed, the most striking example of computation without neurology occurs in the slime mold. Slime molds are perhaps the most alien-like organisms on Earth. The macroscopic slime molds, which are in the minority amongst slime molds, are composed of a syncytium, a collection of what, in other circumstances, would be cells but without cell walls or membranes separating the individual component cells. The result is a single "cell" with thousands of cell nuclei.

Like all motile organisms, the slime mold seeks out food and avoids noxious stimuli. But the way it does so is quite different from the behavior of animals. The slime mold extends itself in a pseudopod-like fashion in many directions simultaneously. Those extensions that encounter food are reinforced while those that contact something unpleasant are diminished.

This is a familiar tune for us. The strategy of creating a multitude of branches and then paring back those not in use is reminiscent of apoptosis. The difference in the case of slime molds is that the nonbeneficial extensions do not die off as would be the case in "preprogrammed cell death" but rather are resorbed into the body of the slime mold. The pseudopodia that have found food are not promoted as are synaptic weights but rather physically reinforced and grow in size.

This strategy of extension and retraction, reinforcement and diminution is the substrate on which complex computations can occur. In 2010, Tero, Tagaki, and collaborators published an article in the premiere journal, Science. In their study, the authors deposited food sources on a plate in a scaled representation of Tokyo and surrounding cities. They inoculated the plate with a slime mold and let it ramify, extend, and contract until it reached a stable net-like branching pattern. Examination of this pattern demonstrated it to have a remarkable similarity to the actual rail network connecting these cities in Japan. In a paper the following year, the authors described "Traffic optimization in railroad networks using an algorithm mimicking an amoeba-like organism, Physarum plasmodium." The year after, Vincenzo Bonifaci, Kurt Mehlhorn, and Girish Varma published a paper in which they described the Physarum slime mold as revealing "an algorithm developed by evolution over millions of years."

And what of us? Do our bodies perform computation? Without a doubt, we are capable of great feats of computation through our brains and other elements of the central and peripheral nervous system. Interestingly, this nervous system is not the only one we possess. A 100-million neuron nervous system resides in our gastrointestinal tract. Called the ENS or enteric nervous system, this collection performs computation and communication necessary to digestion and may play a role in anxiety and depression.

But I have digressed. The quest of this chapter is to identify biological systems of computation that do not involve neurology. Again, our bodies are an excellent example of this. A system exists within our bodies that assesses environmental changes and performs what cannot be described as anything other than very complex computation. When confronted with a pathogen, our immune system performs a detailed structural analysis and then manufactures nanomachines to target precisely these invaders. The immune system, as we call it, is really a collection of different systems, some of which arose very early in our evolution.

Organisms as simple as bacteria produce *defensins*, small polypeptides (i.e. proteins) that combat viruses, fungi, and other bacteria. In a broader sense, the development of antibiotic resistance can be thought of as a type of biological computation, albeit one that plays out intergenerationally.

The cognate in our mammalian bodies is called the innate immune system and, while it demonstrates efficacy in combating pathogens, demonstrates little flexibility. The masters of flexibility, of tailored response to specific invaders are the elements of our adaptive immune system.

When confronted with a novel virus, our adaptive immune system crafts antibodies that fit lock-and-key with surface proteins of the pathogen, identifying it and marking it for destruction. The manner of genesis of these antibodies is not material to our discussion here, but it is worthwhile to note that it occurs without neurology.

The adaptive immune system ingests information, stores the information, and responds to the information. In Markov's terms, it is *determined* in that it gathers information in the form of conformational models of viral proteins. It is *applicable* in that it stores this new information, and it is this information storage that is the reason vaccination works. And it is *effective* in that it generates a generally predictable response that affects the milieu of the stimulus by creating antibodies that ultimately result in the decrease of viral population. By Markov's criteria, indeed by any reasonable criteria, the immune system is performing computation.

When we discuss concepts like computation or intelligence, it is too easy for us to tie these ideas to particular physical structures like computers or brains. It is important to abstract these concepts because it helps us to recognize them in places where we would not have otherwise looked. One of the Alan Turing's great contributions is his conception of a machine capable of performing any computation. In 1936, Turing described the idea of a machine capable of writing a character on a piece of tape and then either moving the tape forward or backward, changing its configuration in accord with the character on the tape. He demonstrated that such a simple machine was capable of performing any computation, of carrying out any program no matter how complex.

The tape on which the character is written is a concept rather than necessarily being an actual piece of tape. It could be a strand of RNA or knots on a string or any other means of storing rewritable information. By abstracting what has become called a Universal Turing Machine, we can see its presence in biology without neurons and in inanimate systems without circuity.

I have intentionally excluded from this chapter a discussion of an almost limitless source of computation within biological systems, the interaction between proteins each of which runs its own electrostatic, chemical, or mechanical program. But unlike the slime mold which can perform remarkable computation as a unitary organism, proteins perform their work only through interaction. They act in the aggregate.

The idea of computation through the aggregate interaction of individual units is exactly what we see in what biologists describe as super-organisms. These are colonial agents that do not reproduce as individuals but only as colonies, that accomplish tasks as groups that no individual members could themselves perform, that compute as populations as if all were parts of a single organism distributed over many bodies. These are the hives and are the subject of the next chapter of our story.

Chapter 14: Superorganisms

Fresh bread is one of life's great pleasures. I've maintained sourdough starters many times, only to doom them to neglect and malnutrition. Inevitably, I return to levain husbandry because fresh bread is a wonder. What's your go-to? Salted butter? What about a nice cheese and garlic spread?

Cheese and garlic spread turns out to be a great deal older than I would have thought. Virgil, author of the Aeneid, wrote a poem that was nominally about moretum, a Roman garlicky herb and cheese spread but actually about a bitter farmer. Nonetheless, instructions for moretum preparation are given:

The reeking garlic with the pestle breaks,

Then everything he equally doth rub

I' th' mingled juice. His hand in circles move:

Till by degrees they one by one do lose

Their proper powers, and out of many comes

A single colour, not entirely green

The verse is written in Latin and the phrase "out of many comes a single colour" is written as "color est e pluribus unus." E pluribus unus, or e pluribus unum, is a catchy phrase and appeared regularly on the title page of Gentleman's Magazine, an early 18th century monthly for the educated man interested in commodities and in Latin poetry.

Six years after the founding of Gentleman's Magazine, Pierre-Eugène Ducimetière was born in Geneva. Having immigrated to New York before the founding of the United States and modifying his name to Pierre Eugene du Simitiere, he had the distinction of forming the first American museum of natural history from his personal collections and, more importantly for our story, suggesting the phrase "E Pluribus Unum" (from many, one) for the Great Seal of the United States, in part because the phrase contains 13 letters, symbolizing the original 13 colonies. This also served as the unofficial motto of the United States until 1956 when "In God We Trust" was adopted by Congress.

E Pluribus Unum was selected, of course, as a metaphor for the United States, one nation from a multitude of colonies. In this 14th chapter, we explore the idea of a superorganism, a unitary entity itself composed of smaller entities. And, as has been the case in all complex systems we have examined, we will see that the superorganism acts differently from any of its component members.

James Hutton, a late 18th century Scottish geologist, is credited with having coined the term superorganism in 1789 as a means to describe the Earth as a single being. Our modern understanding of superorganism as a colony of animals acting as a whole was introduced in 1911 by William Morton Wheeler, a Harvard myrmecologist. One of the defining features of a superorganism is that it demonstrates reproductive division of labor. That is to say, the majority of individuals do not reproduce but rather assign that task to a queen or some other equivalent.

But before we consider what it means to be a superorganism, we need to ask what it means to be an individual. Male and female organisms are different from each other, by definition. That's not to say that animals cannot change gender during the course of their lives and indeed this is seen in clownfish and oysters. In fact, some animals seemed to express no gender until the time of reproduction arrived.

Sigmund Freud is justifiably famous for his work on what has become psychiatry. But before examining humans, Freud spent a great deal of time examining eels. As a student at the University of Vienna, Freud sought to solve the problem of determining gender in European eels, a group of fish that do not demonstrate physical manifestations of gender until they have matured. Freud's research consisted of dissecting eels to definitively determine gender. Having spent a great deal of time looking for anatomical evidence of male gender, he wrote, "but in vain, all the eels which I cut open are of the fairer sex." Indeed, so discouraged was he that he abandoned the field and turned, more fruitfully, to the study of the mind.

In addition to males, females, and hermaphroditic organisms, some animals, even vertebrates, can reproduce through parthenogenesis. This is asexual reproduction in which a female organism, without insemination, can reproduce. Notably, female Komodo dragons have been documented to lay eggs that hatch into viable young without the participation of a male Komodo dragon in the fertilization process. Komodo dragons, which look like dinosaurs but are not, are the largest extant lizard. Turkeys, which don't look like dinosaurs but are, have also been documented to reproduce parthenogenically, although this seems to be a relatively infrequent occurrence.

Sexual reproduction requires two individuals to contribute genetic information to the offspring. But let's examine the word individual in this context. One animal that seems to blur the concept of individuality is the anglerfish. These are frightening looking animals with enormous mouths lined with myriad sharp teeth. Their most distinctive morphologic characteristic is a long thin polyp called an illicium that extends from the head to some distance in front of the mouth. At the end of this polyp is a bulb, the esca, that in some species demonstrates bioluminescence, that is to say, like fireflies it can glow of its own chemical power. This bulb serves as a lure to draw smaller fish on which the angler fish feeds.

The angler fish that I have described is the female of the species. The male is a diminutive fish only about a five-hundred thousandth the mass of the female. Mating, as it were, takes place when the male adheres to the female and then is absorbed into the female's body. Nearly all of the male's body atrophies, leaving only his testes inextricably attached, indeed incorporated, into the female's body. Genetically, this remnant of the male is distinct from the female's body in which it resides. But morphologically, that is to say by any external appearance or behavior, what is left of the male has become part of a unitary organism that is predominantly female. I pose to you a sort of inverse ship of Theseus question. After mating, is this one organism or two? How many individuals are there?

This sort of fusion, this loss of morphologic individuation, is irreversible in vertebrates but not so in microscopic organisms. We discussed in the last chapter route optimization in slime molds. But I left something out. This giant multinucleated syncytium that is the slime mold does not arise as a single body. Rather, individual amoeba-like cells under environmental stress will fuse to form this single "individual". The process is sort of reversible in that when food is found and the environment is less stressful, the slime mold will produce fruiting bodies that manifest as individual amoeba. Given the multinucleated syncytium that is the slime mold, it is more a philosophic than a biological question as to whether these individual amoebae are the same ones that fused to form the slime mold.

Slime molds represent a case of unicellular organisms fusing to form what can either be thought of as a multicellular organism or more appropriately as a multinucleated organism. There are genuinely multicellular organisms that can fuse to form a larger individual. We see this in the case of Hydra, a multicellular polyp that is best known for its ability to regenerate when dissected into multiple pieces. Split a hydra in half and, in short order, you will have two independent hydra.

What is less well known is that adjacent hydra, if injured, can fuse to form a single organism. This merger does not represent a new unitary organism and is more a symbiosis of two organisms in one fused body. However, the genetic identity of the individual cells is maintained in this new chimera.

In the case of slime molds and of hydra, the organism that arises from aggregation is not what we would call a hive. A hive is a true superorganism in that the individual organisms that comprise the hive are physically separated from each other in a way that is not true of slime molds and hydra. But if a hive is simply a collection of organisms interacting or even cooperating as a colony, is there any distinction between what we are calling a superorganism and a conventional community?

A theme upon which we have touched several times is the concept of emergence: that the presence of a behavior of the aggregate is different from the behavior of its individual agents. Put more concretely, do ant colonies exhibit behaviors not found in any individual ants? At first flush, this seems an easy question to answer. All we need do is compare the behavior of individual ants to that of the colony. However, ant colonies demonstrate a division of labor that manifests in physical differences between different ants based upon their roles.

Perhaps to understand individuality among colonial organisms like ants, it is helpful to first examine the counter argument: that ants are, in fact, individuals. Different ants have different types of roles. There are foragers, soldiers, and even nurses. But how is this different from our own society in which different people have different occupations? The distinction among ants is that the difference between different roles is not only behavioral but also physiological. A soldier's mandibles are so enlarged that it cannot even feed itself and must be groomed by workers. A replete ant is a living storage vessel that devotes the majority of its own abdomen to serve as a container for honeydew, a food storage medium for the colony. This is not merely fat to provide a caloric gas tank for a particular ant, but rather food for the community. These morphologically dedicated ants literally cannot live alone. One might even argue, indeed I would, that they make sense only as components of a superorganism.

On reflection, though, there is a serious problem that a superorganism has that an integrated individual like you or me does not have. Our components, our cells, communicate to each other directly or via the circulatory or nervous system. But there's no physical connection between different ants. How do they coordinate their behavior?

In your body, hormones trigger instruction sets within your cells. A program is launched when a cell comes in contact with a particular hormone. The hormone itself does not carry the instructions, but rather a cue to launch an already stored program. The hormone is external to the cell, and so it is not such a leap to think that it might be possible to employ such triggers that are external to the ant itself. The term for a system of external triggers to modify internal instructions is stigmergy, coined in 1959 by Pierre-Paul Grassé, a French zoologist, from the Greek stigma (mark) and ergon (work).

Stigmergy may take one of two forms. Marker-based stigmergy is the type most familiar to us. Individual ants may lay down chemical signals onto their environment and in doing so create a pheromone trail. We can think of this trail as a map written upon the environment itself. The idea seems alien, but if you remember the story of Hansel and Gretel, their behavior was no different. By laying down crumbs and pebbles, they were also writing a map onto the environment itself. When Theseus combated the minotaur, he mapped the path out of the maze by laying down a string as he progressed through its twists and turns. This string also represents an example of marker-based stigmergy at our scale.

In a broader sense, stigmergy is a strategy of employing the environment as a medium to store memory.

A more abstract sort of stigmergy is sematectonic stigmergy. Here, physical changes to the environment trigger programming stored within individual organisms. When a termite mound is being created, individual termites produce columns from material they process. This is program A. However, once one column reaches a threshold height, the surrounding termites abandon their individual columns and coalesce, joining forces with the termite that has built the threshold column. This change in behavior represents the execution of program B, triggered by the observation by individual termites of a single mound that has crossed a particular threshold.

Program B is initiated by the change in the environment that is the threshold column. No hormone is produced, no chemical trail is laid down. But it is still the case that the action of one termite influences the behavior of others.

It is hard to think of an example of sematectonic stigmergy in human society. One approximation is the erection of a stake to guide construction of an oil well. In this context, information is gathered by one specialized set of workers who produce and measure seismic waves to map rock structures kilometers underground. The stake that is erected at the optimal drilling site by one set of specialists serves as a sematectonic cue for another set of workers to erect the larger oil drilling platform.

However, this example of really very primitive sematectonic stigmergy does not begin to capture the breadth of computation that can occur in a stigmergic system.

To glimpse the power and subtlety of stigmergy in the real world, let us turn our attention to the beautiful symmetry of the beehive with its matrix of hexagonal cells. If we think back to Chapter 3 and our discussion about fractals and iterative generation of complicated structure, it would seem that this would be a reasonable pathway for creating the beehive's hexagonal lattice. But on reflection, this would of course be impossible because such fractal construction would consist of creating a single enormous hexagon, and then recursively dividing it into smaller hexagons. This is not the way beehives are built!

Instead, a bee will recognize the arrangement of walls at the edge of the hive and its perception of this arrangement triggers an algorithm within the bee to create the next wall at the appropriate 120° angle. The bees do not lay down a pheromone trail to signal other workers as ants do. But the construction that they create nonetheless serves as an environmental signal in a stigmergic system. The hexagonal design of the hive is the result of an optimal packing algorithm over a planar surface. But, of course, optimization is not calculated at the level of the hive: it has instead resulted from selective pressure.

This clustering of cells in a hive is common to other Hymenoptera, the order including bees, wasps, and ants. But in the case of paper wasps of the genus Polistes, the construction problem is not so straightforward. These wasps create a comb pattern with a convex, sphere-like shape rather than the flat pattern of the beehive. That does not sound like it would make much of a difference, but if hexagons are packed into the curved profile of this hive, they don't arrange evenly, and the honeycomb would be irregular and lobed. The real nests don't look like this. They're beautiful and symmetrical, and the way the wasps achieve this is by introducing a stochastic element into their construction. The same stigmergic cues, that of the observation by the wasp of partially constructed hexagonal cells, trigger a program in the wasp to build the next set of walls. However, unlike the entirely deterministic construction algorithm executed by the bees, the wasp algorithm has a random element. The conventional construction of the next wall at a 120° angle, which occurs all of the time in the case of the bees, is executed only about 55% of the time with the wasps. There is an element of randomness to the wasps' pattern to organize the next cell differently, and the result is a uniform natural-appearing comb. This is stigmergy with an underlying element of randomness, a further algorithmic complication, but one that is entirely biological. The not-entirely-determinative process that is wasps' behavior is not a flaw or even a compromise. It is a feature, a strength that improved the final product. This is noise as an enhancement.

We have seen that stigmergic cues can take the form of pheromone trails and even of structures built by individual members. We're going to extend this concept a bit further in our exploration of the bees' waggle dance. In work conducted in the 1940s, Karl von Frisch discovered that the waggle dance performed by bees conveys information about the location of food sources. This work on communication among insects was so significant that von Frisch received the Nobel Prize in 1973.

Von Frisch demonstrated that the waggle dance itself conveyed information about the location of a food source, and that this information triggered behavior: the observing bees flew to the indicated area to forage for nectar. In this setting, the arrangement of the bees on their "dance floor" acts as a stigmergic cue. Not only are the bees not laying down any pheromones, they are constructing the triggering architecture by arranging themselves.

This work was elaborated upon in 1994 by Thomas Seeley, demonstrating a temporal component, as well as a spatial one. The waggle dance consists of a series of movements by an individual bee, following which the dancing bee returns to its original position and repeats the dance. However, Seeley found that the bees did not quite return to the starting position they had occupied in the previous performance, but rather deviated by just under a millimeter in the direction of the food source. Extending upon this work, David Dormagen noted that the observing bees surrounding the dancer would recenter themselves around the dancing bee in its new position, and the whole ensemble would drift.

This drift is significant for three reasons. First of all, the drift is not directionless, but itself points to the location of the food source. Second, the drift serves to separate groups of bees on the dance floor. Bees surrounding the drifting dancing bee grow further and further from the center of the floor. Simultaneously, other bees observing other dancers drift further in their own direction. Bees pulled to the east side of the dance floor will naturally encounter more dancing bees indicating food sources to the east, reinforcing the message of the original dancer. Finally, and most importantly for our purposes, this drift represents a temporal stigmergic cue.

Thus we have observed four types of stigmergy: marker-based stigmergy, as is the case with pheromone trails laid down by ants; sematectonic stigmergy, as in the towers constructed by termites; stigmergy by arrangement of individuals, as was elucidated by von Frisch; and a sort of fluid stigmergy in which the cue is not so much an object, but the change in an object's position over time.

I have described stigmergy as a system, in which a cue initiates an algorithm intrinsic to an organism. We understand that brains are capable of demonstrating algorithmic work, but what of organisms that have no neurology? Can bacteria demonstrate stigmergy?

Myxococcus are predatory soil bacteria. As a Myxococcus bacterium makes its way through the soil, it deposits a chemical trail. Other Myxococcus bacteria then follow this trail in what is a clear marker-based stigmergic system. The bacterial behavior is, of course, devoid of any neurology. A sort of computation occurs with memory resident in the environment rather than in the bacteria themselves. In this way, Myxococcus represents the entire Hansel and Gretel story being both Hansel and the witch!

In this chapter, we examined what it means to be an individual. We saw that the boundary between being an individual and being a component of an organism is not distinct, but rather forms a spectrum, including organisms like hydra and slime molds that can transform from a community to a single large organism and then fracture back into a community. We introduced the idea of a superorganism and saw how stigmergy acts as a surrogate to the nervous system of a conventional organism like you and me. We marveled at the fact that the spatial arrangement of bees on a dance floor can initiate behavior in the observers of the dance, and we saw that bacteria can store memory in their external environment. And finally, we all agreed, beyond the shadow of a doubt, that fresh warm bread is the physical incarnation of joy.

www.ingramcontent.com/pod-product-compliance
Lightning Source LLC
Chambersburg PA
CBHW051309250726
48656CB00004B/1567

* 9 7 9 8 9 9 5 4 2 3 6 1 4 *